Anonymous

Medical facts and observations

Vol. 6

Anonymous

Medical facts and observations
Vol. 6

ISBN/EAN: 9783337713485

Printed in Europe, USA, Canada, Australia, Japan

Cover: Foto ©berggeist007 / pixelio.de

More available books at **www.hansebooks.com**

MEDICAL FACTS

AND

OBSERVATIONS.

VOLUME THE SIXTH.

LONDON:

PRINTED FOR J. JOHNSON, N° 72, ST. PAUL'S CHURCH YARD.

M.DCC.XCV.

CONTENTS.

I

Being

DIRECTIONS TO THE BINDER.

Plate the First, the references to which are explained in pages 92 and 116, may be placed at page 92; and plate the second at page 121.

MEDICAL FACTS

AND

OBSERVATIONS.

I. *Obſervations on the Uſe of Arſenic in the Inter-*
mittent Fevers of a tropical Climate; to which
is prefixed an Account of the Weather, at Sierra
Leone, during .the Seaſon in whih ſuch Fevers
are moſt prevalent. By Thomas Maſterman
Winterbottom, *M. D. Phyſician to the Set-*
tlement at Sierra Leone.

A S arſenic, though of late years frequently
and ſuccefsfully uſed in England for the
cure of intermittent fevers, has not, to my know-
ledge, been hitherto employed in a tropical climate; ſome account of its uſe in Africa, with
the hiſtories of a few of the caſes in which it
was exhibited, will not, I hope, be altogether
unacceptable.

It may be proper however to premife a fhort account of the weather at Sierra Leone during the feafon in which intermittents are moft prevalent.

The year may be divided into the rainy, tornado, and dry feafons. The rains on this part of the coaft commonly fet in about the end of May, or beginning of June ; and continue, more or lefs violently, until the beginning or middle of September: they are then fucceeded by tornadoes, which continue until the end of November. It muft be obferved, however, that the rains are not only carried off by tornadoes, but alfo brought on by them; and that the tornadoes preceding the rains are, in general, lefs regular than thofe which terminate them. The dry feafon continues from December until May, though fhowers of rain fometimes occur during the dry months.

In 1792, the rains commenced about the end, of May, and continued for fome time to be very heavy; from the middle of July, however, until the laft week of Auguft, there were frequent intervals of fair weather, twelve hours of rain being generally followed by twenty-four or thirty hours of fair weather, with fometimes a bright fun. During this period the thermometer at noon ufually ftood at from 78° to 80°.

The

The laſt week of Auguſt and firſt week of September were remarkable for an almoſt inceſſant rain, which was for the moſt part ſmall and drizzly, though it ſometimes fell in heavy ſhowers; the air at the ſame time felt cold and raw, particularly in the evenings and mornings, when a thick fog covered the hills. The thermometer at noon was from 77° to 80°.

On the 7th of September a tornado came on, which returned on the 10th, 15th, 16th, 18th, 19th, 21ſt, 22d, 24th, 26th, 28th, and 30th.

On the 8th, 11th, 12th, 25th, and 29th, the ſhowers of rain were frequent.

On the 9th, 11th, 14th, and 23d, thunder and lightning occurred during ſome part of the day. The 9th, 13th, 15th, 17th, 24th, and 26th were ſultry and almoſt calm. During the continuance of the rains, the winds chiefly blew from between the ſouth and weſt points, but moſt frequently from the ſouth-weſt, whence alſo the heavieſt rain came.

As ſoon as the tornadoes appeared, the ſea and land breezes had a more regular ſucceſſion; the ſea breeze uſually began from the north-weſt about eight or nine A. M., and towards ſunſet drew round to the weſt: the land breeze then ſetting in from the eaſt or ſouth-eaſt, con-

B 2   tinued

tinued to blow all night and during the early part of the morning.

Towards the end of the month the thermometer generally ftood at 82° at noon, the atmofphere being lefs hazy, and the air cool.

The month of October was throughout attended with regular fea and land breezes; the atmofphere was free from haze, but fometimes overcaft with clouds during the day; the whole of the month was cool and agreeable, though the thermometer at noon generally ftood at 82°, and on the 29th at 84°.

A tornado occurred every night, or early in the morning, from the 1ft to the 18th inclufively, frequently attended with heavy rain for fome hours, and with much thunder and lightning. During the remainder of the month the tornadoes became lefs frequent, occurring only on the 19th, 21ft, 23d, 25th, 27th, 28th, and 29th. The 1ft, 17th, and 24th were fultry. On the 24th it was calm all day. On the 3d there was much thunder and lightning. On the 7th, 15th, 18th, 21ft, and 30th, frequent fhowers of rain fell. The tornado on the 17th came from the fouth-weft which is uncommon. The tornado on the 2d was not followed by rain. The 26th was remarkably hazy all day.

-The

The lightning was extremely vivid on the 28th, appearing in long ftreams or chains of fire.

The month of November was much warmer than the preceding one, the thermometer at noon being from 82° to 84°. On the 11th it rofe to 85°. It was on the 5th at 75°. There was continued rain till noon, when the fky became clear, the day calm and fultry. The atmofphere during the greateft part of the month was clouded and hazy, at leaft the tops of the hills were covered with haze during fome part of the day. The fea and land breezes continued to blow very frefh, but the mornings were frequently calm and fultry till near ten A. M. On the 28th it was calm all day. Tornadoes occurred on the 2d, 8th, 10th, 12th, 13th, 16th, 19th, and 25th. The 5th, 17th, and 23d were rainy. The 5th, 11th, 14th, 18th, and 28th, were fultry, with a little wind.

In December alfo the fky was generally hazy and clouded; the fea and land breezes were pretty frefh during their continuance, but the mornings were for the moft part calm, the fea breeze not fetting in till near ten A. M.; the evenings alfo were clofe and fultry from fun-fet till late at night.

A tornado came on, the morning of the 7th,

followed

followed by much rain, thunder, and lightning; but it cleared up before noon; a heavy shower fell in the afternoon of the same day.

The clearest days this month were the 3d, 9th, 13th, 18th, 24th, and 25th.

On the 5th, 8th, 14th, 15th, and 22d, gentle showers fell: on the 8th there was much thunder and lightning. The weather was sultry, with little wind, on the 1st, 3d, 14th, 19th, 22d, and 27th. The 14th and 27th were calm days. The land wind blew all day on the 13th, and the south-west and south-south west winds on the 2d, 30th, and 31st days. The thermometer at eight A. M. usually stood at from 77° to 80°; on the 13th at 75°, and on the 26th at 81°: at noon it was from 81° to 84°; at eight P. M. from 78° to 80°.

The remittent fever which during the months of June, July, and August, had very generally prevailed here, and had raged with great violence, began to abate in the month of September. Early in the month, this disease had not only become less frequent, but also more mild in its symptoms, gradually changing into the form of an intermittent. Towards the end of the month it became very rare, the cases which occurred being chiefly among the whites, especially

pecially thofe lately arrived in the country; or others who had been irregular and intemperate during the courfe of preceding intermittent complaints.

In the months of October, November, and December, intermittents were fo prevalent, that fcarcely a family in the fettlement, although the whole number was nearly 400, remained perfectly free from them. They generally obferved the quotidian and tertian type; there were, however, a few inftances of double tertians. Moft of the above cafes were fo mild, particularly among the men, as not to prevent them from following their different occupations, except during the time of the paroxyfm. But in fome inftances, the daily recurrence of the difeafe, the long continuance of the paroxyfm, and a poor diet, confifting chiefly of falted meats, rice, caffada, &c. reduced the patients to a ftate of great debility, and infenfibly laid the foundation of long and tedious complaints. The greateft fufferers from intermittents were thofe who had previoufly laboured under remittent fevers, and had not yet recovered their ftrength; alfo perfons of delicate and irritable habits, children, and women giving fuck.

In every inftance where the bark was taken

in due quantities, and perfifted in for a proper length of time, the paroxyfm was fpeedily checked, and the danger of a relapfe effectually prevented; nor did the patient fuffer thofe ill effects which ufually occur where the difeafe has continued long, and been left to itfelf. Few, however, of the common people could be prevailed upon to take the bark in any form; and even thofe who took enough of it to obviate the return of a fingle paroxyfm, would feldom continue it a fufficient length of time to eradicate the difeafe. Thefe confiderations, joined to an apprehenfion that ferious and alarming confequences might enfue from frequent relapfes, determined me to try the effects of the mineral folution, according to the plan recommended by Dr. Fowler *. The fear of difordering the bowels, and inducing dyfenteric fymptoms, rendered me at firft very · cautious in its ufe; but on finding, after repeated trials, that no ill effects were produced by its exhibition, I was encouraged to employ it more generally. The fuccefs with which it was attended will appear from the following detail of cafes :

* Medical Reports of the Effects of Arfenic in the Cure of Agues, &c. 8vo. London, 1786.

CASE

CASE I.

October 4.—S. Peters, a black, aged four years, is affected every day, about noon, with coldnefs and violent fhiverings, which continue near an hour, and are then fucceeded by a hot dry fkin, head-ach, and fometimes vomiting. The paroxyfm is terminated in the evening by a copious perfpiration. In the abfence of the fit he makes no complaint, but appears languid and weak, and has little appetite. A confiderable degree of hardnefs is felt on the left fide, with a tumour projecting below the cartilages of the falfe ribs. He was ordered to take four drops of the mineral folution three times a day.

5. Had no cold fit yefterday at the ufual time, but appeared heavy and uneafy; no ficknefs or griping was occafioned by the drops.

8. Has had no return of the paroxyfm fince the 3d. No griping nor any fenfible effect has been produced by the medicine.

The folution was now omitted, and he took, on the 9th, four grains of calomel. This child had no relapfe, and has continued fince to
enjoy

enjoy good health, although the tumour in the fide did not wholly d.fappear till the beginning of the year 1793.

·CASE II.

October 4.—Hannah Peters, a black, aged thirty-fix years, has been for two months paft affected with an intermittent fever ; at prefent a paroxyfm comes on every day at noon. During the hot fit, fhe has a confiderable pain of the head, efpecially over the eyes, which continues till evening, and is gradually abated by the fweat which then breaks out. Her ftrength and appetite are much diminifhed.

Capiat folutionis mineralis guttas x. ter die.

6. Had no return of fever yefterday at the ufual time ; but towards evening had a flight cold fit, fucceeded by heat and fweating. The paroxyfm, however, was neither fo fevere, nor of fo long continuance as ufual. She felt a little griping in her bowels.

Repetatur Solutio.

8. Has omitted the folution two days, and has had a return of the hot fit each day at the ufual time, without the preceding cold ftage. She was defired to continue the drops regularly.

16. Has

16. Has taken the folution regularly fince the laft report, during which time fhe has not had the laft return of her ague, nor any pain of the bowels.

Omittatur folutio et capiat Infus. Cort. Anguft. ʒiij teɪ die.

CASE III.

October 10.—David Edmonds, a black, aged forty years, has had every day, for near a month paft, a paroxyfm of ague, attended with a very fevere pain of the head. Of late the fit has only returned every.fecond day, beginning about one o'clock, P. M. In the abfence of the paroxyfm he has no complaint but languor and debility.

Capiat folut. min. guttas x. ter die.

11. Had a flight attack yefterday evening, which did not continue long ; he felt no griping or naufea from the folution.

Repetatur Solutio.

16. Has neglected his medicine for fome days, during which he has miffed the cold fit, but had a pretty fmart hot fit every day, towards evening.

Repetatur Solutio.

20. Has had no return of the cold or hot fit
fince

fince the 16th: he continues the folution without experiencing any difagreeable effect from it.

•

CASE IV.

Octob. 5.—J. Barnes, aged thirty-fix years, of a fair complexion, and florid, with red hair, was attacked with the remittent fever about the end of Auguft laft, from which he recovered by a liberal ufe of the bark ; but foon after, on returning to work, and expofing himfelf too much in the fun, he fuffered a fevere relapfe in the beginning of September. His complaint, however, yielded again to the bark, but left him greatly enfeebled. During the remainder of the month of September, he continued to take from ℥i to ℥ifs of bark every day, and returned to his work. About a week afterwards he was fuddenly feized with a cold fit, followed by a hot ftage and a profufe perfpiration, which left him very weak during the apyrexia. His pulfe is now 100, rather hard and quick : he has a fevere attack every day at noon, attended with vomiting, and, during the hot fit, with a quick and hurried refpiration ; he is hot

·and

and reftlefs till late in the evening, and has then very profufe night fweats.

Capiat folut. min. guttas x. ter die.

Oct. 6. The folution did not difagree with him. He had a flight return of the paroxyfm yefterday.

Repetatur Solutio.

8. Has had no return of the fit, nor felt any fenfible effect from the medicine. He perfpired much at night; has great debility and languor, with little appetite.

10. The fymptoms are nearly as before; he did not reft well, but had no return of the paroxyfm.

Capiat opii gr. ij h. s. Repetatur Solutio.

12 The folution was yefterday omitted; he refted better with the pill: in other refpects finds no alteration.

13. Had a return of the paroxyfm yefterday; the cold ftage lafted half an hour, the hot ftage about two hours. He was much relieved by the opium, and fweated very profufely after it.

14. Had another flight fit yefterday evening, the cold ftage being very fhort; he fweated much: does not recover his ftrength or appetite. As he could not be prevailed upon to take the bark again, I directed that four ounces of the

following

following infusion should be taken three times
a day:

℞. Corticis Angusturæ ℥j Cremor. Tart. ʒij Aquæ
pur. ℔ifs.

By this plan his appetite became better, and
he regained his strength in some degree; but
in a week or ten days he relapsed into his former state, having every day an ague fit, which
was, however, relieved by two grains of opium, taken at the commencement of the cold
stage. He now began to take the bark to the
amount of ℥ifs a day, which finally put a stop
to the ague; notwithstanding, he recovered his
strength so slowly, that it was thought necessary,
six weeks afterward, to send him to England
for the effectual restoration of his health.

CASE V.

October 14.—A. Richardson, a black, aged
forty years, since her recovery from a remittent
fever in August last, has continued in a very
debilitated state, and for some time past has
been affected with an intermittent fever, the
cold fit of which comes on daily at four o'clock,
P. M. is very severe, and of long duration.
Much

Much pain of the head, and frequent vomiting attend the hot fit, which continues the greateft part of the night, and is fucceeded towards morning by a flight partial fweat: fhe remains very weak till the commencement of the next paroxyfm; her appetite is much impaired; her body open.

Capiat folut. min. guttas x. ter die.

16. Has taken the folution two days, and has had no appearance of the ague, except a little uneafinefs and yawning about the time of its ufual attack. No fenfible effect is produced by the medicine.

Repetatur Solutio.

17. Had a return of the paroxyfm yefterday; the cold fit was fhort, but fevere; the hot fit was alfo violent, and terminated by a profufe perfpiration; after which, however, fhe appeared more eafy and compofed than ufual. She complained of no griping or naufea from the medicine.

Repetatur Solutio. ·

24. Has had no return of the paroxyfm fince the 17th, nor any fymptoms of its approach. She continues ftill very weak, and has little appetite.

Omittatur folut. Capiat infuf. gent. c. ʒij ter die.

28. Has had no return of the fit. She begins

gins to recover her ftrength and appetite.

Repetatur Infuf.

CASE VI.

Nov. 2.—Mary Bowler, aged forty years, a black, has been for fix weeks affected with a tertian ague; the cold fit is fevere; the hot fit, which is very violent, and attended with great pain of the head, generally continues all night, and fometimes part of the next day, without any fweating ftage. She is much debilitated, but has a tolerable appetite.

Capiat folut. min. guttas x. ter die.

4. Had a return of the paroxyfm yefterday, after the third dofe of the folution. The fit returned at the ufual period, and in the fame manner as before. No fenfible effect was pro-duced by the medicine.

Repetatur Solutio.

8. Has had no return of the cold fit fince the 3d; the hot fit occurred about the ufual time, but it was fhorter and much lefs fevere than ordinary.

Repetatur Solutio.

12. Has had no return of the paroxyfm; fhe complains of a little griping in her bowels, and continues ftill weak.

Omittatur folut. min. Capiat inf. gent. c. ℥ij ter die.

I   20. She

20. She makes no complaint, and has nearly recovered her health and fpirits.

Repetatur Infus. Gent. c.

CASE VII.

Nov. 1.—E. Perth, a black, aged forty-five years, has been for near fix weeks paft affećted with an irregular intermittent, which moſt commonly follows the tertian type. The cold fit is fevere, and very uncertain in the time of its attack and in its duration. In the hot fit ſhe complains of exceſſive pain of the head, efpecially over her eyes, and of great pain of the back. The hot ſtage generally continues all night, feldom terminating by regular fweats: it is followed by much laſſitude and uneafinefs through the enfuing day. Her ſtrength is greatly impaired, her appetite bad; and ſhe is very coſtive.

Capiat ſtatim Sal. cathart. amar. ʒi. Cras incipiat fumere Sol. min. guttas x. ter die.

6. After taking three dofes of the medicine, ſhe had a return of the paroxyfm on the 3d, but thought the cold fit later in its approach than ufual, and ſhorter. The hot fit continued

Vol. VI.   C   throug^u

through a great part of the night, but the pain of the head was much lefs fevere. She has had no return of the paroxyfm fince, and feels only a little griping from the medicine.

℞ Repetatur Solutio.

10. Has had no return of the paroxyfm fince the 3d. She complains only of debility and want of appetite.

Omittatur Solut. Capiat Infus. Gent. c. ℨij ter die.

14. Begins to recover her ftrength ; her appetite is alfo better.

Repetatur Infus. Gent. c.

CASE VIII.

Octob. 3.—Ann Bowler, a black, aged fourteen years, has been, for fome weeks paft, affected with an irregular tertian, which is fometimes, but not generally, preceded by a cold ftage. The hot ftage continues during the greater part of the day, and feldom terminates by fweating. Her body is open; her appetite much impaired.

Capiat Solut. min. guttas viii: ter die.

10. The folution has now been taken for a week, during which time fhe has had no return

of

of the ague, nor has felt any naufea or griping from the medicine. No complaint remains but debility.

CASE IX.

October 4.—Dinah Lawrence, a black, aged forty-four years, is every other day, about fix o'clock P. M., feized with a fevere cold fit, followed by great heat and violent pain of the head, efpecially over the eyes, which fymptoms continue through the whole night, and are not fucceeded by any regular fweating ftage ; fhe is coftive, and much debilitated ; fhe has had this complaint near three months.

Capiat ftatim Sal. cath. am. ʒvi ; et cras Solut. min. guttas x. ter die.

10. Has had no return of the fit fince fhe began to take the folution ; fhe finds no difagreeable effect from it : is ftill coftive.

Repetantur Sal cathart. et Solut. min. ut antea.

14. Feels no complaint but what proceeds from debility; her appetite is better; fhe was a little griped by the medicine.

Omittatur Solut. min. Capiat Infus. Gent. c. ʒij ter die.

CASE

CASE X.

Sept. 24.—Jane Armftrong, of a fair complexion, aged thirty years, is feized every day, at eleven o'clock A. M., with a head-ach fo violent as to produce frequent fhrieking and continual moaning. The pain chiefly affects the crown and one fide of the head; it is in general preceded by a cold ftage, though flight, and of fhort duration. The hot fit, which is not very violent, continues till night, when it abates along with the pain; but is not entirely removed till morning: the paroxyfm is ufually terminated by a profufe perfpiration. The patient is naturally of a delicate conftitution, and has of late been much reduced by the remittent fever, from which fhe recovered very flowly.

Capiat Opii gr. iij et Tart. emet. gr. ¼ ingruente paroxyfmo.

25. The head-ach was almoft entirely removed within half an hour after taking the pill; the paroxyfm terminated alfo more fpeedily than ufual. Being very coftive, fhe was ordered to take half an ounce of purging falt the following morning.

26. The

26. The falt operated gently; fhe had a very violent return of head-ach at the ufual time, which was relieved by the opium taken alone.

October 4.—She refufes to take the bark: fhe has every day had a return of head-ach at the ufual time, which was however removed by the opium.

Capiat Solut. min. guttas x. ter die; et repetatur Opium fub initium paroxyfmi.

10. Has had no return of the paroxyfm fince fhe began the folution; feels no inconvenience from its ufe, but a flight diarrhœa, without any pain.

Repetatur Solutio.

14. Has had no return of the head-ach; fhe fweats much at night; is very weak, and has no appetite.

Omittatur Solut. Capiat Infus. Cort. Anguft. ʒiij ter die.

This woman has never had a return of the paroxyfm, though a twelvemonth has now elapfed fince the laft report. She gradually recovered her ftrength by the ufe of tonic remedies.

CASE XI.

Sept. 12.—Jeffe George, a black, aged twenty

years,

years, was yefterday afternoon feized with a fevere cold fit of an ague, which continued upwards of two hours, and was fucceeded by great
heat, fevere pain of the head, naufea, pains
all over his body, more efpecially in the back
and loins, great reftlefsnefs, and anxiety. Towards morning a general but not profufe perfpiration took place; the feverity of the head-
ach at the fame time abated, and all the other
fymptoms wholly difappeared: he has much
thirft; his fkin is cool; his pulfe 72, and foft.

Capiat Solut. min. guttas x. ter die.

13. He had a return of the paroxyfm laft
night, at eight o'clock, four hours later than
the former one. The cold fit, though very fevere, did not continue long; the hot fit was
ftrong; the head-ach lefs violent. He had a
very profufe perfpiration this morning. His
fkin is now cool and moift, and his tongue clean;
but fome pain ftill remains over the orbits of the
eyes; he complains of thirft, and is coftive.

Capiat Sal. cathart. am. Ʒi—Repetatur Solut. min.

14. The head-ach continued yefterday till
the afternoon, and then went off; the falts were
not taken till this morning. He refted well laft
night, and makes no complaint but of debility.

Repetatur Solutio.

15. He

₁₅. ₃ₑ continued free from complaint yefterday, till towards evening, when he became
hot and feverifh ; and after a very uneafy night,
he, this morning, at eight o'clock, had a fevere
cold fit, attended with violent head-ach, which
lafted near an hour. Two grains of opium,
taken at this time, brought on a fweat, and terminated the paroxyfm.

Repetatur Solutio.

16. He flept well laft night, and feels no
complaint but from debility. He has omitted
the drops this day.

Repetatur cras Solutio.

17. Has had no return of the paroxyfm ; he
feels no complaint but a flight griping from the
folution.

R̥. Tinct. Opii et Solut. min. āā ʒij ɱ. capiat guttas **xx.**
ter die.

20. He has had no return of the paroxyfm
fince the 15th. At that time he probably
brought it on by having expofed himfelf the night
before to the damp evening air in his fhirt. He
feels no griping, or ficknefs, from the drops,
which he ftill takes. He returned to his work
this day.

C 4   CASE

CASE XII.

Auguſt 12, 1793.—Mr. T——, a European, of a dark complexion, with black hair, was ſuddenly ſeized, two days ago, with an acute pain of the head, chiefly over the orbits of the eyes, attended with naufea and vomiting. Theſe ſymptoms were ſoon followed by great heat and reſtleſsneſs, which continued through the whole night, and yielded in the morning to a profuſe perſpiration. On the 11th he was free from complaint; walked about, and ate heartily. In the evening, however, he was ſeized with a very ſevere ſhivering fit, which continued near two hours, and was ſucceeded by great heat and reſtleſsneſs, by ſevere pain above the eyes, and bilious vomiting. He was again relieved in the morning by a copious perſpiration. At ten o'clock, A. M. his ſkin was ſtill hotter than natural, and his pulſe rather quick; in other reſpects he appeared free from complaint.

Capiat Solut. min. guttas x. ter die.

13. The firſt doſe of the ſolution yeſterday produced vomiting; the ſecond gave him three ſtools;

ftools; the laft had no particular effect. He
paffed an eafy night, without feeling any fymptom of the fit, except a general uneafinefs,
which, however, foon went off. He complains
this morning of flight pain over his forehead.

Repetatur Solutio.

14. The medicine again produced ficknefs,
and a flight diarrhœa, though he only took two
dofes of it. He remained well till two o'clock,
P. M.; he then became very hot, and had a
fevere return of the head-ach, attended with
naufea and vomiting. The heat, pain, and
reftlefsnefs continued till this morning, when a
copious perfpiration took place, with which he
is yet affected.

At ten o'clock A. M. his pulfe is 130; his fkin
pretty cool; his head-ach almoft gone; his
tongue fomewhat furred. He complains of thirft,
and of flight pain of his bowels, with a fenfation of numbnefs about the umbilicus

Omittatur Solut. Capiat pulv. Cort. Peruv. ʒi fecunda
quaque hora.

At fix o'clock, P. M. he has a very flight head-
ach, with a fenfe of weight in the forehead; his
eyes are more prominent and brighter than ufual.
He has taken two dofes of bark fince noon,
the firft of which produced vomiting; he has
had

had one ftool to-day; his urine is very high coloured; pulfe 130, foft, and lefs quick than in the morning.

Repetatur Cort. et capiat h. s. Tinct. Opii et Vin. Antim. āā guttas xxx.

15th. Ten o'clock, A.M—he has had a good night; fome pain ftill remains over his eyes, but it is lefs fevere; his fkin is rather hot, but moift; pulfe 112; his tongue dry and white; his urine high coloured, with a light cloud fufpended in it. He complains much of thirft and fever, and of a pain in his back. He has taken, fince yefterday noon, ʒifs of Peruvian bark.

Repetantur Cortex, Tinct. Opii, et Vin. Antim.

16. He paffed an eafy night, and enjoyed fome refrefhing fleep; he complains only of a flight pain over his eyes, and is able to fit up. He had two ftools in the night; his tongue is cleaner, but ftill dry; pulfe 104 and foft, but eafily quickened by the leaft exertion. His urine is not fo high coloured, and exhibits a flocculent cloud. He took ʒifs of bark between ten o'clock, A. M. yefterday, and fix o'clock this morning.

Repetantur Cortex, Tinct. Opii, et Vin. Antim.

17. He was much griped yefterday by drinking fome cyder; has no complaint this morning

ing but from weaknefs. His pulfe is 104, and foft; his tongue clean and moift. His urine is much paler than before, and has a kind of gelatinous ftriated cloud fufpended in it.

The fame medicines were repeated.

18. He feems much better in every refpect; his appetite is returning; his pulfe 90, and foft.

He continued the bark a few days longer, and had no return of complaint.

CASE XIII.

October 4.—Ann and Eliz. Davis, blacks, the former five, the latter fix years old, have been for fome time paft affected with quotidian agues. The cold fit comes on at four o'clock, P. M.; is very fevere, and frequently attended with vomiting. The hot fit ufually continues the whole night, being attended with great reft-leffnefs, anxiety, and acute pain over the eyes; but is feldom fucceeded by a regular fweating ftage. Their appetite and ftrength are much impaired.

Capiant Solut. min. guttas vj. ter die.

9. Each of them had a return of the cold fit

on

on the 4th, after the third dofe of the folution. They have fince had no return.

Repetatur Solutio.

11. There has not been any appearance of the paroxyfm, nor any difagreeable effect from the medicine.

CASE XIV.

John Oliver, a black, aged five years, who was affected nearly in the fame manner as the two laft patients, began, on the 16th of Auguft, to take four drops of the folution three times a day.

23. He had a return of the fit on the 16th, 17th, and 18th, but it commenced every day later, was lefs fevere, and of fhorter duration. Since the 18th he has had no fit, although the folution was difcontinued. A flight tumefaction of the face has been obferved for two days paft, but is at prefent fubfiding. He felt no naufea or pain from the medicine.

Dec. 10.—Mary Jones, a black, aged thirty-six years, about three months ago was affected with a remittent fever, from which she recovered very slowly, and has since continued in a state of great debility. She has of late been subject to violent pains in the bowels, attended with diarrhœa. During the last month she has had a regular tertian ague, the cold fit of which begins generally at sun-set, but is not very severe, nor of long continuance. The hot fit is long and severe, being attended with violent head-ach, intense thirst, and great restlessness. These symptoms are not terminated by a regular sweating stage; and have often no remission till the middle of the following day. She is feeble, and much emaciated.

Capiat Solut. min. guttas x. ter die ; et Opii. gr. ij. sub accessionem paroxysmi.

12. The hot fit was much relieved by the opium; the paroxysm was shorter, and the head-ach less severe. She is very costive.

Repetatur Solut. min. et capiat Sal. cathart. ʒss mane.

15. Con-

15. Continues the folution without feeling any fenfible effect from it. She has had no cold fit or head-ach during the two laft paroxyfms. The hot fit was much lefs violent and of fhorter duration than formerly.

Repetatur Solutio.

18. Has had no return of the fit, nor any appearance of it fince the laft report; nor does fhe perceive any naufea or griping from the folution. Her appetite is ftill much impaired.

Repetatur Solutio. Capiat Inf. Cort. Anguft. ʒij ter die.

22. There has been no return of the paroxyfm. She finds her ftrength and appetite much increafed by the infufion.

The ufe of the folution was difcontinued.

CASE XVI.

Feb. 1.—John Jones, a European, of a fallow complexion, aged twenty-eight years, is affected in the afternoon, every other day, with a violent cold fit, attended with rigors, and fucceeded by a regular hot fit and fweating. Until within a few days, he has been able to do his duty on fhip-board as a feaman; but the paroxyfm returns now with fo much violence,

I   as

as to confine him to his hammock. He has taken a large quantity of Peruvian bark at different
times, which has never failed to prevent the next
return of the paroxyfm; he has always, however, had a relapfe in a few days, through in-
temperance, and expofure to the night air.

Capiat Solut. min. guttas x. ter die.

8. Has taken the folution without perceiving
any fenfible effect from it. The paroxyfm returns as ufual, but, as he fays, with much lefs
violence.

Repetatur Solutio.

15. The paroxyfm returns as ufual, but is
fhorter and lefs fevere. Through miftake, he
has taken the folution only before the attack of
each paroxyfm.

Repetatur Solutio; et capiat guttas x. ter die.

20. He has had no return of the paroxyfm
fince he took the folution as directed, and feels
no naufea or griping from it.

He continued the medicine a few days longer,
and was reftored to perfect health.

CASE

CASE XVII.

Feb. 1.—Ann Wicks, a mulatto, aged forty years, has been for a month paſt affected, every other day, with a violent cold fit, attended with rigors, and ſucceeded by great heat. She has alſo a ſevere pain over the forehead, and on one ſide of the head, extending to the neck and ſhoulder of the ſame ſide. There is much ſtiff-neſs and pain in moving the neck during the intermiſſion. The cold ſtage commences about five o'clock, P. M. and continues near an hour. The hot fit does not terminate before morning, and is ſeldom ſucceeded by a regular ſweating ſtage. She is much debilitated by the long con-tinuance of the complaint, and has lately given ſuck to a young child. Her appetite is alſo greatly impaired.

Capiat Solut. min. guttas viij ter die.

4. She has had no return of the cold fit. The hot fit continued only part of the night, and was unattended with head-ach or any other dif-treſſing ſymptom.

Repetatur Solutio.

12. She

12. She has had no return of the paroxyfm, and feels no ill effects from the folution. Her ftrength is fomewhat increafed, but her appetite is ftill bad.

Omittatur Solutio. Capiat Inf. Gent. c. ʒifs bis die.

CASE XVIII.

Mrs. D. a delicate woman, of a fair complexion, aged twenty-four years, in the month of Auguft laft had a mifcarriage, from which fhe recovered without much trouble, and enjoyed a tolerable ftate of health till the beginning of October, when fhe was feized with the common remittent fever of the place. From this complaint fhe alfo recovered within a fortnight, by taking largely of the bark in powder and decoction. About the end of the month, however, fhe fuffered a relapfe, and made a very flow progrefs towards recovery; her ftomach being only able to retain the bark in the form of a decoction. She laboured under great debility, very profufe night fweats, and frequent hectic flufhings during the day, with lofs of appetite, and general tremors on ufing the leaft exercife. Thefe fymptoms were at length

confide-

confiderably alleviated by the infufion of An-
guftura bark, elixir of vitriol, and other tonics.

Dec. 15. Yefterday, at fix o'clock, P.M.
fhe had a cold fit, with rigors, which lafted near
half an hour, and was fucceeded by a hot
fit, attended with great pain of the head, naufea, vomiting, and reftleffnefs, which continued through the whole night; towards morning fhe was relieved by a partial fweat, but remained very weak and languid.

16. Yefterday, at the fame hour, fhe had a
return of the paroxyfm, the fymptoms of which
were mitigated by an opiate taken foon after its
commencement: fhe had a copious perfpiration
during the night, and feems free from complaint this morning.

18. Had a return of the fit on the 16th and
17th, but was relieved as before by an opiate.
She refufes to take bark.

Capiat Solut. min. guttas viij. ter die.

20. She has had no return of the cold fit fince
the 18th. The hot fit was much fhorter and
lefs fevere. She experiences no inconvenience
from the medicine.

Repetatur Solutio.

22. She has had no return of the paroxyfm,
but feels a flight pain in her bowels.

Capiat ftatim Tinct. Opii guttas xx. et fp. lav. c. ʒfs.

Repetatur Solut. min.

24. The

24. The pain of her bowels was removed by the opiate; she has had no return of the paroxyſm; reſts pretty well during the night, but ſweats much towards morning.

Omittatur Solutio ; et capiat Infuſ. Gent. c. ʒi ter die.

30. Her ſtrength is returning. Her appetite is good, and ſhe has had no return of the paroxyſm.

This lady continued to enjoy a good ſtate of health, till the 20th of March, 1793, when ſhe was affected with a diarrhœa, attended with acute pain in her bowels, chiefly about the umbilicus. She was ſoon relieved from theſe complaints by an opiate, and a few powders, conſiſting of the colombo root joined to an aromatic : but on the 25th, ſhe had a return of an intermittent fever, the cold fit of which was very ſevere. It began at ſix o'clock in the evening, continued near two hours, and was followed by a hot fit, which laſted all night, terminating towards the morning in a ſlight perſpiration, and leaving her low and weak the remainder of the day.

28. She refuſed yeſterday to take an opiate on the approach of the cold fit, having on former occaſions found her head diſagreeably affected by it. The paroxyſm proved very ſevere : the

D 2　　　　　　hot

hot fit continued all night, and was fucceeded by partial fweats about the head and neck. She is very weak this morning, and complains of a great pain of the head and back; of lownefs of fpirits and general uneafinefs.

Capiat Solut. min. guttas viij. ter die, ex Infuf. Cort. Anguftur. cyatho.

30. The folution did not difagree with her in any refpect; fhe had a cold fit laft night, but it was much lefs fevere than ufual: fhe is alfo in better fpirits to-day.

Repetatur Solutio.

April 1. There was no cold fit yefterday; but fhe had a hot fit, which continued all night, and terminated in a very profufe perfpiration. Her fpirits are much revived; fhe is confiderably ftronger, and has a better appetite.

Repetatur Solutio.

6. She continues the folution without feeling any inconvenience from it; and has had no return of the fit, or night-fweats, fince the 1ft: her appetite at prefent is good.

Repetatur Solutio.

8. She has had no fit, and recovers her ftrength gradually. No naufea or griping has ever been produced by the folution.

Omittatur Solutio. Capiat pulv. rad. colomb. gr. xv. ter die.

CASE

Feb. 1, 1793—Mrs. H. of a fair complexion, aged twenty-four years, during the months of September and October laft, had two feveral attacks of the remittent fever, from which fhe recovered fpeedily by means of the bark : fince that time fhe has continued in a very weak irritable ftate, fubject to pains of the bowels, and to frequent though flight returns of a febrile complaint, which continued only for a day or two, and commonly yielded to an opiate. On the 27th of January fhe had a cold fit at eight o'clock in the morning ; this was fucceeded, in about an hour, by a burning heat of the fkin, with flufhing of the face, great reftleffnefs, and fevere pain of the forehead. Her eyes, at the fame time, appeared bright and prominent ; fhe complained alfo of a fenfe of heat in them, and was unable to bear the light. In the evening, a copious perfpiration enfued, and confiderably alleviated the fymptoms ; fhe had, however, a flight head-ach through the whole night : the

D 3   fit

fit has returned every morning at the fame time for the laft four days.

Feb. 2. The paroxyfm appeared this morning as ufual, with a fevere cold fit and head-ach, but was rendered much fhorter and lefs diftreffing by an opiate draught taken foon after its acceffion.

Capiat Solut. min. guttas viij. ter die.

5. She had a flight return of the cold fit this morning, with a little head-ach, but the paroxyfm was of fhort duration.

Repetatur Solutio.

6. She has had no cold fit to-day, nor any pain of the head; the hot fit returned at the ufual time. Her face is much flufhed, and her fkin hot, but with lefs anxiety and reftleffnefs than heretofore: fhe finds no inconvenience from the folution. The opiate was not taken to-day.

Repetatur Solutio.

10. She has had no return of the paroxyfm, nor has felt the flighteft fymptom of its approach fince the 6th; fhe complains only of a flight pain or uneafinefs in her ftomach. Her appetite ftill continues weak.

Omittatur Solut. min. Capiat tinct. opii guttas xx. ftatim.

14. She begins to recover her ftrength and appetite;

appetite ; the pain of the ſtomach was immediately removed by the opiate.

All the patients whoſe caſes are here related,
have continued to enjoy good health ſince cured
by the ſolution; and though ſeveral months
have now elapſed, none of them have experienced the leaſt unpleaſant ſymptom which
could be attributed to that remedy. The women
continued to labour under a ſuppreſſion of the ca-
tamenia, until their ſtrength was entirely reſtored.

Mrs. H. (Caſe XIX.) though enjoying a good
ſtate of health, had no appearance of them till
the middle of Auguſt laſt, when they flowed
for ſeveral days rather profuſely.

In Caſe IV. I had little proſpect of ſucceſs
from the uſe of the ſolution, the child having
become very weak and irritable by frequent re-
lapſes : but as he had for a length of time
taken the bark in large doſes without any effect,
I was induced to try the mineral ſolution, with
a view of checking the returns of the paroxyſm,
hoping afterwards to complete the cure by the
bark ; which might prove more effectual after
its uſe had been ſuſpended a few days.

D 4   In

In Cafes I. X. XIII. and XIV. there was an evident enlargement of the fpleen, forming a projection below the cartilages of the ribs. In Cafe X. it was fo large as to extend nearly as low as the crifta of the os ilium. After the ague had ceafed, the patient continued to ufe corroborant medicines, taking at the fame time fmall dofes of calomel, but without any fenfible effect on the tumor; it yet remains nearly in the fame ftate, not, however, caufing much uneafinefs. In Cafes XIII. and XIV. as the patients fpeedily regained their health after the ague had ceafed, and felt no uneafinefs from the enlargement of the fpleen, I did not think it proper to ufe any medicine, excepting a purgative dofe of calomel occafionally, becaufe, in many fimilar cafes, where this medicine had been ufed, even in very fmall dofes, a falivation was very foon excited, the tumor not being at all affected by it, whereas the patient was rendered extremely weak and irritable. The only inftance of tumefaction which could with any probability be referred to the ufe of the folution, was Cafe XIV. in which, however, it proved fo flight, as fcarcely to deferve notice.

In order to give the mineral folution a fairer trial, I avoided, in many inftances, making

ufe

ufe of two very powerful means ufually employed for the purpofe of diminifhing the violence of the paroxyfm, and which frequently
indeed put a total ftop to it; I mean, opium
and emetics: when two grains of opium are
given a fhort time before the paroxyfm is
expected, it feldom fails to bring the fit to a
fpeedy termination by a profufe fweat; and
generally relieves the violent pain of the head,
which is fo diftreffing during the hot fit, as in
Cafes X. and XV. The recurrence of the paroxyfm being once obviated, I have found that
a full dofe of opium at night affords more com-
fortable reft, and more certainly prevents the
folution from affecting the bowels, than when
the tincture of it is added to the mineral folution; a mixture of this kind always becomes
turbid, and the opium is partly feparated.

    Intermittents partake much of the nature of
remittents, and the two difeafes have a very
uncertain boundary; whenever, therefore, the
intermiffions are imperfect and indiftinct, the
exhibition of an emetic is attended with moft
beneficial effects. In many inftances this practice puts a temporary ftop to the returns of the
fit, and in every cafe confiderably diminifhes
its violence. The proper time of giving an

emetic

emetic, is about two hours before the paroxyfm is expected; and the beft mode is to employ a folution of tartarized antimony in divided dofes, at intervals of eight or ten minutes, until full vomiting be produced. When the patient has vomited a few times, and his ftomach is a little fettled, a more moderate dofe of the antimonial folution, joined to a full dofe of opium, feldom fails to produce a copious perfpiration before the attack of the cold fit. This method generally fucceeds in preventing the immediate recurrence of the paroxyfm : but in thofe cafes where the intermittent has continued long, and feems to return by the power of habit, it will be proper to repeat the emetic once or twice more before the time when the paroxyfms are expected.

I think it proper here to obferve, that antimonials, in the naufeating dofes in which they are frequently given during the remiffion or apyrexia, with a view of procuring a more perfect folution of the difeafe, are feldom found adequate to the purpofe; on the contrary, the continued action of fo powerful a ftimulus, in general, produces a correfpondent ftate of debility, and relaxes the mufcular fibres of the fto-

2   mach

mach fo much, that neither food nor medicine can be properly retained.

The remittent fever is, in many cafes, very mild; whence the remiffion has often been miftaken for an intermiffion. This miftake is more liable to be made when the remittent fever is preceded by an evident and fevere cold ftage at each return of the paroxyfm, and is followed by a regular hot, and fweating ftage. The remittent may, however, be diftinguifhed from the intermittent fever; 1ft, by a flight pain which remains fixed in the forehead, or over the orbits of the eyes, during the apyrexia; 2dly, by the pulfe, which, though not more frequent than in health, yet retains a degree of quicknefs or fharpnefs through the whole of the remiffion; 3dly, by the ftate of the fkin, which, though moift, feels hotter than natural. In fuch cafes I have not found the mineral folution fo fuccefsful as in thofe where the intermiffion was complete; for which reafon it feems moft prudent to place our fole dependance upon the bark, as in Cafes IV. and XII. Sometimes, however, when the patient could not be prevailed upon to take the bark in proper dofes, I have found much advantage from joining it with the mineral folution, by which means a fmaller quan-

tity

tity of bark will anfwer the intended purpofe.
But whenever immediate danger prefents itfelf,
or is to be apprehended from a continuance of
the fever, the baik, given in large dofes, is the
only medicine to be depended on.

The mineral folution ufually fails in fome
irregular cafes, which at firft view refemble intermittents, and have been improperly ranked
with them, under the denomination of erratic
or anomalous intermittents. A morbid increafe
of irritability appears to be the foundation of
thefe irregular complaints ; they affect prin-
cipally thofe who have been debilitated by frequent attacks of fever, or by lingering difeafes ;
alfo children; and women, more efpecially thofe
who give fuck ; and, in general, perfons of a
weak delicate habit. The fymptoms which
occur in thefe complaints are nearly as follow :
during the afternoon, or towards evening, the
patient becomes uneafy and reftlefs ; his fkin
feels dry, and is hotter than ufual, but without imparting the burning heat ufually obferved
in the hot ftage of intermittents; the pulfe becomes quick, and rather more frequent than
natural; a pain is fometimes felt in the head,
either on the crown, or on the back part of it ;
the thirft is feldom very great; difagreeable
clammi-

clamminefs, however, takes place in the mouth. Thefe fymptoms are fometimes preceded by flight chills running down the back, which, however, when they do occur, are not of long continuance, and never accompanied with violent fhiverings.

In this manner the patient is harraffed during the whole night*, but obtains relief towards morning, when a partial fweat fometimes appears about the head and breaft. Excepting a degree of languor and debility, little or no complaint is felt till the return of evening. The duration of thefe complaints is very uncertain ; they fometimes affect the patient daily for one or more weeks ; at other times abate or difappear for a few days, and then return as before. Whatever increafes the irritability of the body, may be confidered as an occafional caufe of them ; but the moft common as well as moft powerful one is too much fatigue, along with expofure to a hot fun.

In thefe cafes, after evacuating the ftomach and bowels by a gentle emetic or purgative, it is commonly fufficient to exhibit fome tonic, in a form agreeable to the patient's ftomach. The

---

* Hence the denomination of night-fever.

Peruvian

Peruvian bark does not appear to produce any better effects than the other vegetable tonics, as Gentian, Colombo, &c. An infusion of Angustura bark is what I most frequently employ, and find most useful, taking care to prevent the costiveness arising from its use, by giving, at proper intervals, a dose of calomel.

For children, who cannot easily be induced to take bitters, after the previous use of an emetic, a few moderate doses of calomel are commonly sufficient.

Notwithstanding the effects of arsenic appear to be equally as powerful and nearly as certain as those of bark in the cure of intermittent fevers, yet it must be confessed that perfect strength is less speedily recovered when the cure has been accomplished by arsenic alone, than when bark has been employed. This objection to the use of arsenic is of less consequence in cold climates, where, if the ague has not been of long standing, the debility induced by it is seldom very considerable. In tropical countries, however, a few attacks of an intermittent frequently reduce the patient so much, that even when the paroxysm has ceased to return, the extreme debility which remains, is

of

of itfelf fufficiently alarming to demand every attention from the practitioner.

It does not appear improbable that the bark owes its fpecific power, in the cure of remittent and intermittent fevers, to fome peculiar principle in its compofition, which has hitherto eluded the refearches of experimenters, and which they have in vain attempted to imitate by various combinations of bitters and aftrin- • gents. In whatever this peculiar power of the bark may confift, the fame quality appears to be poffeffed by the arfenic in a confiderable degree. Both remedies probably effect the cure of intermittents, by their action upon the fibres of the ftomach, fince they often operate fpeedily, and even in a fmall dofe; but the power of the arfenic feems to ceafe here ; whereas the bark is capable of reftoring tone to the fyftem in general. The fame effect may perhaps be nearly obtained by joining fome tonic medicine to the arfenic. With this view, in many cafes, after the folution had been taken a week or ten days, I difcontinued its ufe, and ordered the patients to take the Infuf. Anguft. Infuf, Gent. c. &c. until their ftrength was completely reftored. It may be found ftill more advantageous to employ

ploy these remedies along with the mineral solution.

Arsenic seems to have been oftener employed
as a medicine in Germany, than in any other
part of Europe; but chiefly by the empirical
class of practitioners, which no doubt prevented its introduction into general use. Many
eminent physicians in Germany, as well as
elsewhere, have, however, spoken highly in
its favour, and occasionally prescribed it. Like
many other active remedies, it has been much
abused by the bold and the ignorant, and has
been given in doses which no man of prudence
would venture to direct; especially as we know
that the same good effects may be obtained by
moderate doses of it, and without the least risk.
The following observations, extracted from a
German work *, will show how extensively this
medicine has been used on the Continent, and how
little caution has been observed in its exhibition.

Dr. Slevogt, Professor of Anatomy at Jena,
in 1700, recommended the use of arsenic, ex-
tolling it as the best, most certain, and safest
cure of intermittents, especially of tertians and
quartans. He employed it in doses of a grain

* Nicolai Recepten und Kurarten. 8vo. Jena, 1780.

or

or a grain and a half mixed with a proper quantity of Theriaca; not only giving it on the days
of the apyrexia, but alfo a fhort time before
the acceffion of each paroxyfm. . He afferts,
that in fifty inftances, two or three dofes were
fufficient to put a total ftop to the difeafe, and
that he never obferved the leaft ill effect from it.

Melchior Friccius * recommends arfenic in
intermittents, and declares he has ufed many
drachms of it in the cure of fuch fevers; but
confeffes that he had often met with relapfes
afterwards.

Lanzonus † quotes a letter from Valifnieri to
one of his friends, written in 1707, in which he
fays the French furgeons were accuftomed to cure
long-continued intermittents with a fmall quantity of arfenic : and he adds, that their remedy
feemed to refemble much the famous aqua del
petefino, which was a ftrong folution of arfenic
boiled in a copper veffel ‡.

* De Virtute Venenorum medica. 8vo. Ulmæ, 1701.—
See alfo London Medical Journal, Vol. VII. p. 194.

† Lanzoni Oper. omn. med. phyf. 4to. Lauf. 1738.
Tom. I. p. 68.

‡ The *Aqua della Toffanina* (fo called from the
inventor), *Aquetta di Napoli, Poudre de Succeffion, Eau Mi-
rable,* &c. were preparations of arfenic frequently ufed as
poifons during the laft century.

Keil * praifes arfenic as a certain and fafe fpecific in intermittents, when prepared and adminiftered in the following manner: half an ounce of white arfenic, finely powdered, is to be put into a glafs, or tea-cup; half an ounce of diftilled vinegar is then to be added, and evaporated over the fire, being conftantly ftirred at the fame time with a wooden fpatula; the fame quantity of vinegar is again to be added and evaporated in like manner. After this procefs has been repeated fix times, the refiduum is finally to be wafhed with warm water, and dried; a drachm of the dry powder is to be made up into fixty pills by means of a fcruple of wafers foftened with water. Previoufly to the ufe of the pills, the patient is to take an emetic compofed of tart. emet. or fulph. aurat. antim. and a little vitriolated tartar, or fome purgative medicine on the morning free from fever: the next day, or only a few hours before the acceffion of the paroxyfm, one of the pills is to be taken fafting, and nothing is to be eaten or drank after it for three or four hours. When this has been repeated three days, during the apyrexia, or a few hours before the

* Anatom. Chirurg. Medicin. Handbuchlein. 8vo. Konigfberg, 1761.

2   attack

attack of the paroxyfm, the fever commonly ceafes. He affirms that this practice has been attended with fuccefs in feveral hundred cafes, when every other remedy had been employed in vain; that he has never obferved the leaft ill effect to accrue from it; but, on the contrary, that thofe who had before looked thin and ill, had become, in confequence of it, fat and ftrong; and that he knew many perfons who had ufed this remedy fifteen or twenty years before, and who continued to enjoy a ftate of perfect health.

Dr. Jacobi * recommends the ufe of arfenic ftrongly in fevers: he directs one part of arfenic and twelve of falt of tartar, to be mixed with 180 parts of water, and boiled till one half has evaporated; when cold, as much frefh water is to be added to it as has been loft by the evaporation, together with a little fpirit of wine. The dofe for adults is twenty-five drops, to be given on the day which is free from fever, at feven A. M., at three, fix, and nine, P. M. Before the ufe of this me-dicine, the primæ viæ muft be evacuated by emetics and purgatives; and the common febri-

* De prudenti ufu Arfenici, fale Alcalico domiti, interno falutari, Differt.—Vide Act. Acad. Elect. Mogunt. Tom. I. p. 216. 8vo. Erford. 1751.

fuge

fuge remedies fhould be ufed for fome time. Dr.
Jacobi obferves that he has employed the above
preparation not only in intermittents, but alfo in
continued fevers, with the greateft fuccefs, and
without ever experiencing any bad effects from it.

Heuermann * fays that arfenic is ufed in Hol-
ftein, at Copenhagen, and fome other places,
as the moft certain remedy for the cure of intermittents; that he has himfelf given it with
conftant fuccefs, in fevers, to patients who were
not able to retain other medicines on the ftomach in a proper quantity; and that two cafes,
wherein frequent relapfes had occurred, were
entirely cured by this remedy. He prepares a
folution of arfenic in the following manner:
half an ounce of white arfenic, and fix ounces
of alkaline falt, are added to ℔ifs of water,
and then evaporated to drynefs. The fame
quantity of water is added a fecond time to
the refiduum, and evaporated to one half,
which is coloured red by a few poppies. Of
this he directs from feven to ten drops to be
taken during the day, beginning immediately
after the paroxyfm is over, and omitting it a
fhort time before the return of the next. If the

* Vermifchte Bemerkungen und Unterfuchungen. Vol. I.
8vo. Copenhagen, 1765.

folution

folution produces vomiting, it is too ftrong, and muft be diluted ; only one dofe is to be given in twenty-four hours, and the patient muft be kept moderately warm, to promote a gentle perfpiration. Expofure to cold, he fays, is as hurtful during the ufe of this as of other febrifuge remedies, as it difturbs Nature in her operations, and retains in the body the noxious matters which fhe is endeavouring to expel. If in the firft three or four days after the ufe of thefe drops, the fever does not ceafe, he re-commends that the fame dofe fhould be repeated twice a day, which commonly proves fufficient. The ill confequences which have been obferved after the ufe of arfenic, as palfy, trembling of the limbs, blindnefs, deafnefs, &c. he afcribes to the improper preparation and imprudent ufe of it; afferting, that it is a fafe remedy when properly prepared.

In the Ephemerid. Acad. Nat. Curiof.* arfenic is alfo celebrated as an infallible fpecific for in-termittents. Three or four grains of powdered white arfenic are directed to be put into a fmall uncovered glafs with a proper quantity of water, and placed upon the fire till a folution takes place, when it is to be well ftirred up and given to the patient: the fever, we are in-

* Dec. II. Ann. III. p. 132.

formed,

formed, is by this means certainly prevented from returning. The patient fhould eat nothing for twelve hours before; but a quarter of an hour after having taken the medicine, he is allowed a gill of warm water, in which a quantity of butter is diffolved, together with the yolk of an egg; after which, nothing more is to be given for fome hours. There generally follows a confiderable degree of uneafinefs, and a profufe fweat; and by thefe means, it is faid, every intermittent, even a quartan, may be readily cured. Two other formulæ are given in the fame work[*], and recommended as highly ufeful in the cure of intermittents, viz.

R. Tart. crud. ʒi. Arfen. cryft. ʒfs. Pip. long. ʒfs. Lap. prunell. ʒifs. Specif. febrifug. Crollii ʒiij. M.

The dofe is from gr. v. to Ɔfs.

The other is

R. Arfen. alb. gr. v. Lap. prunell. vel Nitri depur. gr. xv. M. pro una dofi.

Profeffor Ackermann [†] relates, that in Paufa, a town of Saxony, a furgeon's family had been poffeffed for more than a century of a fecret remedy againft melancholy, which confifted of two grains of arfenic mixed with a drachm or more of white fugar, to be taken

[*] Dec. II. Ann. V. p. 474.

[†] Neues Magazin für Aerzte. Vol. II. p. 401. 8vo. Leipfic, 1780.

early in the morning, along with a large quantity of mucilaginous d ink. The medicine produced a violent vomiting, fo as to agitate the whole body, which continued not lefs than fix hours; after this, he obfervcs, the patient ufually enjoyed a quiet fleep, and became more rational. The remedy was perfifted in, care being taken that the effects of the firft dofe fhould be completely over before a fecond one was adminiftered. Many repetitions of the medicine were not however requifite, as the difeafe, in general, foon yields to this mode of treatment; the patient was afterwards directed to continue a mucilaginous diet for a few weeks. Profeffor Ackermann examined fome of the patients who had been cured by the furgeon at Paufa, and found that no ill effects had arifen in confequence of it. The fame perfon, it is added, employed arfenic very frequently for the cure of intermittents; he diffolved two grains of arfenic in a pint of water, and gave two, or three table fpoonfuls for a dofe every day; under this treatment the fever feldom recurred more than twice; but he remarked that the patients were longer in recovering their ftrength than when the bark had been ufed.

Profeffor Ackermann farther obferves, that another furgeon in the fame place likewife

employed

employed arfenic with great fuccefs; he gave fifteen drops of a folution of arfenic in water, along
with alkaline falt, but the Profeffor had not been
able to afcertain the exact proportions. A dofe
was ordered to be taken as foon as the patient
felt the approach of the fit, and a quantity of
warm tea was to be drank immediately afterwards.
This produced a vomiting, which was encouraged as much as poffible by repeated draughts
of the tea. In this manner, it feems, he had
cured many obftinate agues by two or three dofes
of the folution; and, amongft others, a quartan
which had continued upwards of two years.

From fome of the foregoing narratives, arfenic feems to have been ufed with as little pre-
caution as emetic tartar; and fince it appears,
on good authority, not to have been productive
of bad confequences, even in very large dofes,
we may be induced to lay afide that extreme
anxiety with which we generally prefcribe it;
and may be encouraged to perfift in the ufe of a
remedy which, when prudently adminiftered,
is both fafe and efficacious.

Many of our moft active and approved medicines, as preparations of mercury and antimony, the fquill, foxglove, &c., are capable
of

of producing as violent effects in the constitution, when given in too large a dose, as arsenic
itself. All these medicines met with the same,
if not stronger, opposition when first introduced,
as arsenic does from many at present. It is
well known that antimonial preparations were
declared to be poisonous, and that the use of
them was prohibited by a decree of the faculty of
Physic at Paris in the year 1566; which decree
was not repealed till 1637. We shall cease, however, to wonder at the prejudices formerly entertained against these medicines, when we
consider, that even at the present day similar
objections are made upon the Continent, especially in Germany, to the use of the bark, a
remedy, the reputation of which has been so
fully established by the united testimony of so
many eminent practitioners, supported by almost innumerable experiments.

Mr. Theden, one of the most celebrated sur-
geons in Germany, and Surgeon General to the
Prussian army, in speaking of the treatment of
intermittents, observes *, that when his patients
had previously enjoyed a healthy state of body,
he was generally able to effect a cure in six or

* Unterricht für die Unterwundärzte. 8vo. Berlin, 1793.

eight

[53]

eight weeks. As he entertained the common idea that bark is apt to produce obstructions and enlargement of the viscera, œdematous swellings of the extremities, &c. he cautiously avoided giving this remedy until he had tried every other means. During the first three weeks he employed different medicines, with a view to loosen the morbific matter, and to render it fit for expulsion from the body; he then gave two ounces of bark, in doses of half a drachm, every two hours. After an interval of eight days, during which only bitters were prescribed, he ventured again to exhibit an ounce of the bark, and thus completed a cure. He cautions us against the use of bark whilst the face retains a yellow tinge, or whilst the febrile matter remains in the constitution; he confesses, at the same time, that he has seen œdematous swellings of the lower extremities after agues where no bark had been employed.

Dr. Vogel* is likewise of opinion that many cachectic diseases, particularly obstructions of the viscera, dropsy, jaundice, phthisis, tympanitis, coughs, asthma, hemicranium, deafness, cataract, vertigo, &c. are frequently the con-

* Handbuch der praktischen Arzneywissenschaft. 8vo. Stendal, 1781.

sequences

fequences of an improper treatment of intermittents; more efpecially when the cure has
been attempted by aftringents, arfenic, &c. or
even by an unfeafonable exhibition of the Peruvian bark, whilft the morbific matter ftill
remains in the fyftem.

The objections to the ufe of thefe medicines
are fo vague, that they appear to originate from
popular prejudice and ill-grounded theories,
rather than from any juft practical deductions;
they will therefore have little weight with thofe
who are not contented with bare affertions, but
make actual obfervation and experience the
ftandards of truth.

Having frequently found the moft beneficial
effects from the mineral folution, and having
never obferved any ill confequences to arife from
its ufe, I may prefume to recommend a trial of
it to furgeons practifing in warm climates, and
particularly upon the coaft of Africa.

The high price of bark may fometimes prevent furgeons of fhips from laying in, at their
own expence, fuch a ftock of this valuable
medicine as will enable them to employ it freely
in every cafe which requires its ufe. For notwithftanding the frequent complaints of feveral
refpectable furgeons in the navy, the quantity
of

of bark allowed by government to fhips on foreign ftations, is much too fmall; and moft of the merchant fhips trading to this coaft are ftill more infufficiently provided.

Of the two moft frequent difeafes upon the coaft of Africa, the remittent and intermittent fever, it is certain that the latter, though lefs rapid in its courfe, and apparently lefs dangerous than the former, yet for the moft part occafions that irremediable injury to the conftitution, which fo often befalls Europeans trading upon this coaft. There are few, even of thofe who are faid to be feafoned to the climate by long refidence, who have not fuffered feverely from repeated attacks of intermittents. This in a great meafure arifes from the unhealthy fituation in which they live for the convenience of trade. They generally fix their refidence on the banks of fome river, or narrow creek, whofe oozy fhores, furrounded by mangroves, and excluded from the wholefome breezes, are a conftant fource of miafmata and contagion; to this muft be added the debauched and irregular courfe of life which moft of them lead. Though feafoned to the climate, as they fuppofe, their unhealthy fallow complexions and emaciated bodies,. the frequent hectic flufhings of the
face,

face, swelled legs, &c. attended with obstructions and enlargement of the abdominal viscera, sufficiently indicate to every observer the shattered state of their constitutions. The ague probably still continues to return once a month or oftener, and harrasses them a few days, without being much noticed; for the severity of the disease seems to be considerably abated by its frequent recurrence, though its bad effects in the end are equally certain. As their appetite during the intermission is frequently keen, and even voracious, they flatter themselves that the constitution is not impaired by frequent returns of the disease; many also are negligent, from a confidence in the popular prejudice, that a cold fit shows the absence of danger.

In these cases, therefore, when the bark cannot be procured, or, as more frequently happens, when the patient has conceived a disgust for it, and cannot be prevailed upon to take it in a sufficient quantity, the mineral solution promises to be a safe and effectual substitute for it.

During the last rainy season I have had frequent

quent opportunities of exhibiting the mineral folution in intermittents with the fame good effects as in the preceding year. Out of the number of cafes which occurred in the prefent feafon, I have felected the two following, as being the only inftances of quartans I have met with fince I began to ufe the mineral folution.

CASE XX.

Sept. 11, 1793.—John Thompfon, a mulatto, aged thirty years, was feized, about two months ago, with an ague, which returned every fecond day. After the fecond paroxyfm he took an emetic, and foon after the operation of this, an opiate, which appeared to put a ftop to the difeafe. A month ago he was again feized with cold fhiverings, followed by an increafe of heat, which terminated by a profufe fweat. The fit now returns every fourth day; the cold ftage of which, commencing about noon, is very fevere: the hot ftage continues through the whole night, with violent head-ach, and towards morning is relieved by a profufe fweating.

ing. His appetite is pretty good; his body open.

Capiat Vefp. Antim. Tartar. gr. ij. cu. P. Ipecac. ℈j.
Cras incipiat fumere Sol. min. guttas xij. ter die.

2o. The emetic operated well. He took the folution regularly for four days, and then omitted it, finding no return of his complaint.

3o. He has had no return of the paroxyfm, nor has taken any medicine fince he left off the folution.

CASE XXI.

Sept. 8.—Anne Crankepoor, a black, aged twenty-eight years, has every fourth day, at noon, a fevere cold fit of the ague, which continues near two hours, and is attended with violent rigors and pains of her bones; thefe fymptoms are followed by a hot ftage of long continuance, but which terminates by profufe fweating. She is affected, during the whole paroxyfm, with violent pain of the head, ftomach, and back, which alfo continue through the intermiffion, though with fome abatement. She has taken an emetic and two anodyne draughts without any relief; and

has

has had no ſtool for eight days. Her head-ach
is at preſent very ſevere; her pulſe quick and
frequent; her ſkin hot and dry.

Capiat ſtatim Camphor. gr. x. Tinct. Opii, guttas xxv,
Aq. font. ʒſs. et cras mane Sal. cathart. amar. ʒiſs.
part. vicib.

9. She ſweated profuſely with the draught,
and is much eaſier this morning. Her head-
ach is conſiderably relieved; her pulſe ſoft and
regular. Both doſes of the Sal. cath. amarus
produced vomiting.

Capiat ſtatim Ol. Ricini ʒi.—Repetatur Hauſtus h. s.

10. She could not yeſterday retain the oil
on her ſtomach, nor has yet had a ſtool. She
paſſed an eaſy night, and feels no complaint
this morning, excepting great languor and laſſitude, with a ſenſe of weight and fulneſs in the
abdomen.

Capiat ſtatim Calom. gr. v. Extr. Cathart. Əj.

11. The pills operated gently three times;
her bowels are much eaſier; ſhe feels a ſlight
pain of the head and general uneaſineſs, as if
the fit was approaching.

Incipiat cras ſumere Solut. min. guttas x. ter die.

. 13. The fit returned on the 11th at the uſual
time with great violence. The pain of her
head and ſtomach was alſo very ſevere; ſhe yet
feels

feels fome pain of her ftomach, with great reftleffnefs and uneafinefs. The folution has not been taken till this day.

17. The paroxyfm returned at the ufual time on the 14th, when fhe was affected with very fevere head-ach and pains of the ftomach and back, which ftill continue, being accompanied with great languor. She has taken only five dofes of the folution fince the 13th, and thofe not at regular times. She was very coftive on the 15th, when fhe took

Calom. gr. v. c. Extr. Cathart. gr. xv.

which operated twice. She expects the paroxyfm to-day.

Repetatur Solutio.

18. The paroxyfm did not return yefterday, until fix P. M.; the cold ftage was very fevere, and attended with great pain of the ftomach and head; but thefe fymptoms were much relieved by two grains of opium. She fweated profufely during the night, and feels a flight head-ach and pain of the ftomach this morning, with languor and debility. Her body is open; her pulfe natural.

Repetatur Solutio. Sumantur Opii gr. ij. urg. dolore Ventriculi.

20. She continues ftill weak and languid;

the pain of her ſtomach was wholly removed by the opiate.

Repetatur Solutio.

23. She has had no return of the paroxyſm ſince the 17th, and makes no complaint but of debility; ſhe is, however, able to walk about, and her appetite is ſomewhat better.

Omittatur Solut. min. Capiat Infus. Corticis Anguſt. Ʒiij ter die.

Early in October ſhe had entirely recovered her health and ſtrength.

———

II. *An Account of the good Effects of a Solution of Sal Ammoniac, in Vinegar, employed, as a topical Application, in Caſes of lacerated Wounds. By Mr.* Henry Yates Carter, *Surgeon at Kettley, near Wellington, in Shropſhire.*

IN the ſecond volume of Medical Facts and Obſervations[*], I took occaſion to mention, in a curſory manner, the good effects I had experienced, in lacerated wounds, from a ſolution of ſal ammoniac in vegetable acid, em-

* P. 14.

ployed

ployed as a topical application; and which, in
fuch cafes, I obferved, had feemed to promote
the union of the parts and to moderate the difcharge. As this mode of treatment is very
different from that commonly in ufe, and I
have had occafion to try it in many cafes of bad
compound fracture, and other lacerated wounds,
in which there has been a tendency to fphacelus, I have been induced to make it the fubject
of a diftinct paper, and for this purpofe have
felected the following cafes, from a greater
number, in which I have ufed it; and thefe, I
hope, may be deemed fufficiently interefting
to procure their infertion in a future volume of
the valuable collection above referred to.

CASE I.

A poor man, named Ingram, aged upwards
of eighty years, received an injury on his right
foot, from a carriage paffing over and lacera-
ting it from the inftep to the toes. The wound
had been neglected for fome days, when I was
requefted by a benevolent gentleman in the
neighbourhood to vifit him, and found the foot

fphace-

fphacelated as high as the ancle, and the inflammation apparently extending ftill farther.

I began with fcarifying different parts of the
foot, by which means I gave vent to a confiderable quantity of acrid ichor. The whole foot
was then well covered with lint, continued to
fome diftance above the difeafe, and directed
to be kept conftantly wet with a mixture compofed of half an ounce of crude fal ammoniac
diffolved in a pint of vinegar. Internally he
took the bark in fubftance, liberally, with
opium, as he had a difpofition to diarrhœa.

On the fecond day after this mode of treatment had been adopted, I had the fatisfaction
to find that the inflammation had not fenfibly
increafed, and that the patient felt at intervals
a throbbing, but which, he faid, was not painful, about the ancles. His pulfe, which had
been much quicker, was now at 100.

On the fixth day, a vifible feparation of the
morbid parts was difcoverable, and matter was
perceptible on the verge of the feparating parts;
a fluctuation was felt in feveral parts of the
foot, particularly beneath thofe places that had
been fcarified; and upon making deeper incifions here, we difcovered a collection of good
pus and granulations of new flefh. In the
courfe

courfe of a fortnight, the floughs, having previoufly become loofe, were gradually taken away, and the parts expofed one clear uniform wound. After this the bark was adminiftered lefs frequently, but the ufe of the lotion was continued till the wound was nearly healed, which happened in about two months.

CASE II.

A girl, aged nineteen years, was attacked by a maftiff, and had the mufcles of the thigh and leg, particularly the vaftus externus and gaftrocnemius fo violently lacerated, that the worft confequences were to be expected from the circulation being cut off in the large veffels from the extremity, notwithftanding which fhe loft little or no blood; a circumftance, by the bye, that frequently occurs in lacerated wounds. She fuffered but little pain, although the feparated mufcles of the upper part appeared to be much irritated. The large portions of mufcle yet adhering were cautioufly replaced as near their original fituation as the nature of the cafe would admit; and after the parts had been well

F 3   bathed

bathed with warm vinegar, and due proportions of lint applied round the limb, the whole was encompaffed with a broad roller, applied merely tight enough to retain the dreffings; the limb was then laid in an horizontal pofition, and the preffure taken from the affected part by means of a pillow placed under the lower part of the leg, confiderably below the injury. The whole was then wet with a lotion compofed of half an ounce of crude fal ammoniac diffolved in a pint of vinegar, and ordered to be kept fo con-ftantly.

The firft day fhe was but little fenfible of the application. At night a draught, containing twenty drops of laudanum, was given, and fhe refted well.

On the fecond day I found her pulfe but lit-tle quickened, and her thirft moderate; fhe had perfect feeling in every part of the limb, and complained of an acute fmarting in the wound upon every renewal of the lotion, which continued for a few minutes, and then fhe became eafy. An opening draught was given this morning, and fhe repeated the opiate at night.

On the third day matter feemed to be form-ing,

ing, but there was no appearance of inflammation or fwelling of the limb.

On the fifth day from the receipt of the injury, the bandage was carefully removed, and I had the fatisfaction to find that the mufcles had united, and that the parts of the bone that had been laid bare were covered with new flefh. The difcharge was kindly, and in moderate quantity, and the limb was free from pain. The fame mode of dreffing and the fame applications were continued without alteration during three weeks, at the end of which time the cure was complete.

CASE III.

A young man, aged nineteen years, by a fall of coal in the pit while he was ftooping, was preffed to the ground, and had his thigh broke about four fingers breadth above the patella. The upper part of the bone was forced through the mufcles and into the ground, fo that the hollow of the bone was filled with dirt, and ftripped bare nearly four inches, and the mufcles much lacerated. In this fituation

he

he was brought home, (about a mile) and I then faw him; the wound bled but little.

In this cafe I determined to try the effect of keeping the limb gently extended, nearly at its original length, after taking off fo much of the bone as I fhould find requifite to a complete and exact reduction and to get above the coal flack which had been introduced.

As the bone was fhivered longitudinally, I found it neceffary to take off about three inches of it. This being done, and the wound well cleanfed with warm vinegar and a fmall pro-portion of fpirit of wine, I placed the lower part of the limb as exactly parallel to the other as poffible, and retained it in that pofition by means of proper bolfters on each fide of the limb. An eighteen-tailed bandage having been pre-vioufly laid under the part, the dreffing was made by gently filling the vacancy (the whole fide of the leg *externally* being open) with foft pledgets of lint dipped in the fame folution as that ufed in the preceding cafe, and the ban-dage was then applied as gently as poffible, in order to prevent the flefh from being preffed into the part that the bone ought to have occu-pied; and a fplint applied externally on each fide, merely to give more fteadinefs to the

I

limb,

limb, but without occafioning much preffure. I think it right to mention alfo that the middle tails of the bandage were cut fmaller than the others, and applied in fuch a manner that the wound might be uncovered, in order that the lotion might be applied immediately to the wound, without difturbing any other part.

H· was let blood, and twenty-five drops of tincture of opium were given at night, and the attendant was ftrictly enjoined to keep the part conftantly wet with the folution, except only du·ing the intervals of fleep.

Upon vifiting him the morning after the accident, I found he had had but little fleep, though his limb had given him but little pain, except for about a quarter of an hour after the application of the lotion, after which he faid he had felt the whole leg and foot become fenfibly warmer. The lower part of the limb lay very fteady, exactly in the fituation in which it had been placed; he took this morning three grains of calomel, which procured one ftool.

On the fifth day, including the day of the receipt of the injury, (there having been fome appearance of matter between the folds of the bandage) the dreffings were wholly removed, and the wound was found covered with a well-
concocted

concocted pus in moderate quantity, and with new granulations. The dreffings were continued in the fame manner as before, the whole vacancy being carefully filled with doffils of lint, made as foft as poffible, till the whole was level with the fkin; and over thefe the bandage was appl'ed as before. He continued to repeat the opiate every night, and the calomel occafionally; his appetite was tolerably good, he ufed nearly the fame diet as when in health, and was permitted to drink a fmall quantity of ale.

On the eighth day the dreffings were again removed, and the appearances continued to be favourable. From this time, the weather being warm, the wound was dreffed every day in the fame manner as at firft; and in about eight weeks the callus was completely formed, and had filled up the void fpace, and the wound was reduced to about a quarter of an inch in diameter.

In ten weeks he came down ftairs, and went about on crutches; and in about fixteen weeks from the time he received the injury, he went with a ftick only, and was able to walk nearly two miles. The limb was not quite an inch fhorter than the other; the fmall ulcer continued

nued to difcharge, till a confiderable exfoliation of bone, which gradually made its way out-wards, was extracted, after which the wound foon healed.

CASE IV.

A boy, aged about fifteen years, had the misfortune to flip his hand under the axletree of a water-wheel, which moves at about the dif-tance of two inches and a half from a brick wall or buttrefs fupporting another building; his arm was taken in to the elbow, and the machine performed feveral revolutions on the part before he could be extricated. The flefh was ftripped down on each fide of the thick part of the arm, and the thumb was nearly feparated; but the fingers and hand had fuf-fered but little, and there was no hæmorrhage. The thumb was not taken off, but carefully re-placed, as well as the other mufcular parts that had been feparated; and to the whole wound a large quantity of lint was applied, wet with the folution of fal ammoniac in vinegar. He took twenty drops of tincture of opium at night,
but

but he was very reftlefs, and complained much of his arm.

Second day. The arm had bled in the night, and the dreffings were become ftiff and hard, which rendered it neceffa y to remove them. The difturbance this occafioned produced a degree of inflammation which, I believe, might otherwife have been prevented, and which proved the fource of misfortune. The parts from this time became exceffively painful, and the inflammation extended to the upper part of the arm, and to the fhoulder and fide, as far down as the pectoral mufcle. He was coftive and feverifh, and complained much of thirft. The whole arm was wrapped in a cataplafm made of oatmeal, with equal parts of vinegar and water; and three grains of calomel were immediately given. Two ftools were procured by this medicine; but the pains ftill continued to be very diftreffing to him. His pulfe was at 100.

Third day. The above fymptoms continued; the pulfe was increafed to 110; and he was at times delirious; the upper parts of the arm, fhoulder, and fide, were become of a dark red colour, and were exceedingly tenfe. He had feveral loofe ftools; the arms and fide were dreffed

dreffed as before, with the addition, in the liquid of which the poultice was made, of half an ounce of crude fal ammoniac, and an ounce of fpirit of turpentine. He took half a drachm of Peruvian bark, with fifteen drops of tincture of opium, every third hour; and care was taken to diftil fome of the folution between the dreffings, upon the fhoulder, very often, in fuch a manner that it might make its way to the affected parts.

Fourth day. I found the whole fore arm, from the elbow, completely fphacelated and dry; but the fhoulder and fide were nearly in the fame ftate as yefterday, the inflammation not having increafed; his purging had ceafed; he was not fo thirfty, and his pulfe was at 100; but he complained much of head-ach and wearinefs. Notwithftanding there appeared fome reafon to conclude that his head-ach might, in fome meafure, be occafioned by the quantity of opium he had taken, I continued the ufe of it in the fame dofes; a ftool was procured by means of a clyfter. The ufe of the lotion was continued.

Fifth day. The fymptoms were nearly the fame as yefterday. The fame dreffings and medicines were continued.

Sixth

Sixth day. The pain and tenfion were much leffen.d he had refted tol·ably well, and was fice from thirft; the fhoulder and fide, with a confider:ble part of the upper arm, feemed approach·ng to their natural colour, and the extent of inflammation· was vifibly decreafing. The bark was ftill continued, but without the tincture of opium, inftead of which he took two grains of purified opium at night.

The cataplafm was continued as before for about a week, from this time, when the fhoulder and fide having recovered their original tone, it was changed for one compofed of oatmeal and the folution alone. In a few days matter formed plentifully round the bone in thofe parts where the lacerations had been deep, and large portions of the mufcles were cautioufly removed. The matter formed was of a good confiftence, and moderate in quantity; and the wound was perfectly eafy, except'ng only upon the application of the lotion, and for fome fhort time after. The whole hand dropped off at the wrift; the other parts gradually filled up with good flefh, and are now completely healed.

CASE

CASE V.

A man, aged thirty-fix years, by the fall of a very heavy iron rod perpendicularly upon his foot, upon that part where the fhoe is generally buckled, received a confiderable lacerated wound, by which the tendons were much injured, and the integuments and mufcular flefh were ftripped off from the upper part of the tarfus, and hung in a large loofe flap down the fide of the foot. The wound bled confiderably, and the whole foot, from the violence of the blow, was infenfible. The parts were well cleanfed from the grumous blood with vinegar and water, with a fmall quantity of fpirit of wine, and the loofe flap replaced in the fituation from which it had been torn, and dreffed with pledgets of lint dipped in the folution; and a cataplafm applied of oatmeal and vinegar.

The morning after the injury, upon removing the dreffings, the wound and whole foot were found to have a favourable appearance; but at night he began to complain of a great degree of heat, throbbing, and fenfe of tenfion.

On

On the third day, on removing the dreffings, the whole upper part of the foot appeared to be haftily approaching to a fphacelated ftate. It had loft all fenfibility to the touch, and the inflammation had increafed, though in fo fhort a time, confiderably above the ancle, and to the extremity of the toes. A fenfation of burning heat in the whole foot and leg ftill continued. The parts that were loofe were now removed, and the wound, after having been bathed a confiderable time with a mixture of warm vinegar and water, with a fmall quantity of fal ammoniac previoufly diffolved in it, was dreffed as ufual, the lint being firft well faturated with the lotion; and over the whole a cataplafm was applied as before. A purgative medicine, compofed of four grains of calomel, and five grains of aloes, was given, which operated well. He paffed this day with fomewhat more eafe, and at night took thirty drops of tincture of opium.

Fourth day. He complained of having paffed a very reftlefs night, and that the painful fenfation of burning heat ftill continued; the inflammation went on increafing; his pulfe was at 97, and he had much thirft and flufhing heat. Bark, in the quantity of half a drachm,

was

was given every third hour, and twenty drops of tincture of opium every fixth hour. The fame dreffings were continued, with the poultice; but at night the poultice was omitted, and the dreffings kept wet with the folution alone.

Fifth day. He had refted much better; his thirft was more tolerable, and the heat and other fymptoms were much more moderate; his pulfe was at 90; the inflammation had not in-creafed; and the tenfion about the ancle was leffened. The fame medicines and local appli-cations were continued as laft night. On renew-ing the dreffings in the evening, he complained of having paffed a very painful afternoon, and that the fenfe of heat had been greater. He attri-buted all this to the omiffion of the poultice, which was now, at his earneft requeft, renewed.

Sixth day. In the morning the fymptoms were much increafed, and the inflammation was fpreading, with a violent degree of pain and tenfion, the whole upper part of the foot being in a fphacelated ftate; and the patient com-plained of exceffive pain. The fame dreffings as before were applied, but without the poul-tice, after bathing the parts with warm vine-gar; a broad roller, for the convenience of keeping the parts wet, was gently applied over all the inflamed parts; and as I had a fuf-picion that the increafe of his pain, &c. yefterday, if not wholly, was, in a great meafure, owing rather to a want of due care in keeping the parts conftantly moift, and thus fuffering them to get dry and hard, than to any effect the application could have in producing thofe fymptoms, I paid this day a particular attention to this circumftance, by vifiting him feveral times, to fee that the folution was duly applied; and in a few hours the fymptoms of pain and heat in the whole limb were greatly diminifhed, and continued gradually to abate the whole day His pulfe at night was at 93.

Seventh day. The fymptoms were nearly the fame as yefterday; the inflammation, upon the whole, was rather lefs, but there was no appearance of matter. He had paffed a tolerable night; but his pulfe was ftill at 93. As he was coftive, the purgative medicine was repeated.

Eighth day. He had paft a good night, comparatively fpeaking; the pain in the upper part of the limb (or above the difeafe) was confiderably leffened, and the inflammation was much lefs; a fmall quantity of matter appeared upon the edges of the lacerated parts;

I

his

his pulfe was at 90. He began to complain of
fevere fmarting upon the renewal of the lotion,
and at times infifted on its application being de-
ferred to longer intervals, though when the parts
began to grow dry, the heat and fenfe of ftric-
ture were conftantly renewed.

Ninth day. He had paffed a reftlefs and
painful night; his foot and leg were in much
pain at intervals, but (exclufive of the fmarting pain for a quarter of a hour upon the lotion being applied) he always became much
eafier after the wetting of the parts, which took
place once in about two hours, unlefs fleep
intervened.

From this time the ufe of the lotion was
continued in the fame manner as before, and
he continued alfo to perfevere in the ufe of the
bark and opium; the floughs feparated kindly;
the inflammation went off from the leg and
toes, and a feparation of the difeafed parts took
place at a very little diftance from the edges of
the original injury. The wound difcharged a
well-formed matter, and as the parts beneath
fome of the thickeft floughs granulated, the
latter gradually came away without much pain,
and the whole was healed in ten weeks, except a very fmall ulcer upon the lower part

of

of the Tarfus, through which a fmall exfoliation made its way.

As in the preceding cafes I was careful to obviate the effects of irritation, by keeping the bowels
moderately open, giving occafionally, and fometimes liberally, of opium; and invigorating
the fyftem by means of wine and the Peruvian
bark; it may perhaps be fuggefted, by fome
readers, that the favourable termination of the
cafes I have been relating was due rather to the
internal than external remedies employed; and
that to fubject to a fair and decifive trial this or
any other remedy, no other fhould be employed
at the fame time. This is indeed what I have
done in flighter cafes of laceration, where local
applications only were requifite; and in all fuch
cafes the union of the parts has appeared to
me to be much more fpeedily effected by means
of the lotion, than it is by the ordinary mode
of treatment. And I am able to recollect no
inftance of bad compound fracture, or of lacerated wounds, attended with or threatening
fphacelus, where the warm fomentations and
cataplafms commonly employed in fuch cafes
were

were made ufe of, in which there was any fuch obvioufly good effect from the local treatment, as in the cafes I have been defcribing; notwithftanding there was the fame liberal ufe of opium and Peruvian bark, &c. internally. On the contrary, I have but too often feen the worft effects from fuch cataplafms, &c.; and in one of the above cafes, (Cafe V.), the bad effects of a poultice, applied at the earneft requeft of the patient, were very ftriking, when contrafted with the relief he afterwards experienced from the ufe of the lotion.

III. *Cafe of a difeafed Kidney. By the fame.*

A SEAMAN, forty years old, of a plethoric habit, applied to me at Port Royal, in Jamaica, in 1782, with complaints nearly as follow :

A conftant aching, and fometimes acute pain, about the region of the right kidney, attended with a numbnefs of that fide, and pricking

pains

pains along the urethra, particularly when he paffed his urine; frequent inclination to make water, fomeimes without ability to void any, and never voiding it but in fmall quantity; the urine itfelf being high coloured, depofiting a gritty lateritious fediment, fmelling very ftrong, and forming a film on its furface, which approached to a yellow colour. He complained likewife of a fenfe of fulnefs and heat at the neck of the bladder and about the perinæum, and could get but little reft in any other than an horizontal pofture. He was coftive, and had frequent naufea.

As he had a full pulfe, ten ounces of blood were taken from the arm, and a purging draught was adminiftered; after which he took occafional dofes of a mixture, the principal ingredients of which were diuretic falt and tincture of opium.

In the courfe of two or three days his pain was much alleviated, but the difficulty with which he voided his urine ftill continued.

He now complained of frequent and painful erections, more efpecially when an inclination to make water came on; he had likewife profufe colliquative fweats, and was coftive.

Care was taken to obviate this difpofition to coftive-

coſtiveneſs, by means of purgative medicines
and clyſters. Opium was now more liberally
adminiſtered, and recourſe was occaſionally had ·
to the warm bath. This laſt produced a certain
degree of eaſe while he remained in it, but the
ſenſe of ſtricture about the neck of the bladder
continued, and the quantity of urine he was
able to void ſeemed every day to become leſs,
ſo that at the end of a fortnight it was deemed
neceſſary to make uſe of the catheter, as he
was unable to paſs a ſingle drop of urine without it.

By means of this inſtrument, from four to
ſix ounces of turbid urine were drawn off twice
a day. He had now much fever, and the pain
about the neck of the bladder was become very
acute, and ſeemed to affect him ſpaſmodically,
as well after as previouſly to the introduction of
the catheter. He was likewiſe frequently ſeized
with violent pain, which began in his ſhould-
ers, and proceeded along the right ſide to the
hip.

About a month after the firſt uſe of the catheter, he complained of a pain in the urethra,
near the ſeat of the proſtate gland, particularly when the inſtrument was paſſing; and

at

at times the catheter feemed to meet with fome refiftance at that part.

From this circumftance, together with the continuance of the pain in that and the neighbouring parts, and the frequent difcharge of drops of a mucous confiftence from the urethra, we were inclined to think that the principal feat of the difeafe was in the proftate gland, (efpecially as no appearance of calculus had been obferved), when a frefh fet of fymptoms directed our attention more particularly to the right kidney.

Thefe fymptoms confifted in a pain about the region of that kidney which he had before fcarcely mentioned, but which now (about feven weeks after he firft made his complaints known) was, at times, very fevere. His fhoulders alfo, but particularly the right, were fore, and at intervals acutely painful; the inguinal and axillary glands became fwelled, and fore to the touch; and he complained frequently of a fenfe of coldnefs in the direction of the right ureter, which was fucceeded by a painful inclination to make water.

From thefe circumftances it was fufpected that the right kidney, if not the chief fource of the extraordinary fymptoms I have been

defcribing,

describing, had at least suffered confiderably. He was therefore urged to recollect any external injury he might have received. After a little hesitation he informed us, that about a month previously to his first applying for relief, he had received several violent blows from the end of a large rope acrofs his loins, which for some time had given him confiderable uneasi-nefs. In the courfe of a few days, however, he faid, the pain had gone off, but had returned at intervals; and as · he had suffered much, at different times, from gravel, he had afcribed his prefent complaints to that caufe.

At the time he made known thefe particulars, he was in a very reduced condition; his ftomach was become fo extremely irritable, that it retained but little of what was given to him either of food or medicine; and about a week afterwards he died.

On diffection the urethra was found to be in a healthy ftate, but the proftate gland was a little enlarged. The bladder contained about eight ounces of turbid urine, mixed with a purulent fluid, very offenfive to the fmell. The right ureter was much enlarged, and filled with the fame kind of fœtid matter. The kid-

ney

ney on the fame fide was enlarged nearly to thrice its natural fize, and on being opened was found to be in a ftate of fuppuration, and to contain a confiderable quantity of fœtid pus, fo that the internal fubftance of the kidney was in a great meafure deftroyed.

There was no appearance of calculus; and the other kidney, as well as the reft of the abdominal vifcera, appeared to be in a natural ftate.

It may be doubted, perhaps, whether the affection of the kidney, in this cafe, ought folely to be attributed to the effects of the blows that were inflicted; but allowing the kidney to have been previoufly difeafed (and the complaints the patient had already experienced, and which he attributed to gravel, render it not improbable that it was fo); ftill there can, I think, be no doubt that the fuppurative procefs which took place was haftened, if not immediately occafioned, by external violence. And of fuppuration of the kidnies from external injury, in any refpect fimilar to the prefent, I have been able to meet with no example in books. Different fyftematic writers do indeed enume-

enumerate external contufion among the remote caufes of nephritis, but I do not find, in any of them, an inftance of fuch an affection from fuch a fource; fo that I flatter myfelf the cafe I have related will be thought worthy of being recorded.

It fhows that a frequent inclination, without ability, to make water, is not always occafioned by gravel or calculous concretions; and it affords a ftriking inftance of the influence an organ like the kidney may have upon parts not only contiguous to, but even remote from the feat of difeafe.

IV. *Cafe of a Gun-Shot Wound of the Head. By the fame.*

A HESSIAN grenadier, aged between thirty and forty years, being one of a detachment fent to reduce a fort on the banks of
the

the Delawar, in the act of levelling his piece, received a ball (grape shot) on that part of the os frontis which forms the external canthus of the eye. The ball making its passage through the head, came out under and rather behind the opposite ear, as in the annexed plate *.

What were the immediate effects upon the receipt of the injury I am not able to say, not being immediately upon the spot; but he appeared, when brought to the regimental hospital, to have a perfect recollection of every circumstance that had occurred to him, except only for a short time after he fell. He complained of little pain, and did not appear to have lost so much blood as might have been expected.

The ball being a spent one, had much splintered the cranium, both at its entrance and exit; and was found in the folds of his coat collar.

The wounds being cleansed, and the splinters of bone removed, as far as was practicable, from about the external parts, suitable

* See Plate I. Fig. 1. in which *a* refers to the entrance of the ball, and *b* to the part where it passed out.

dressings

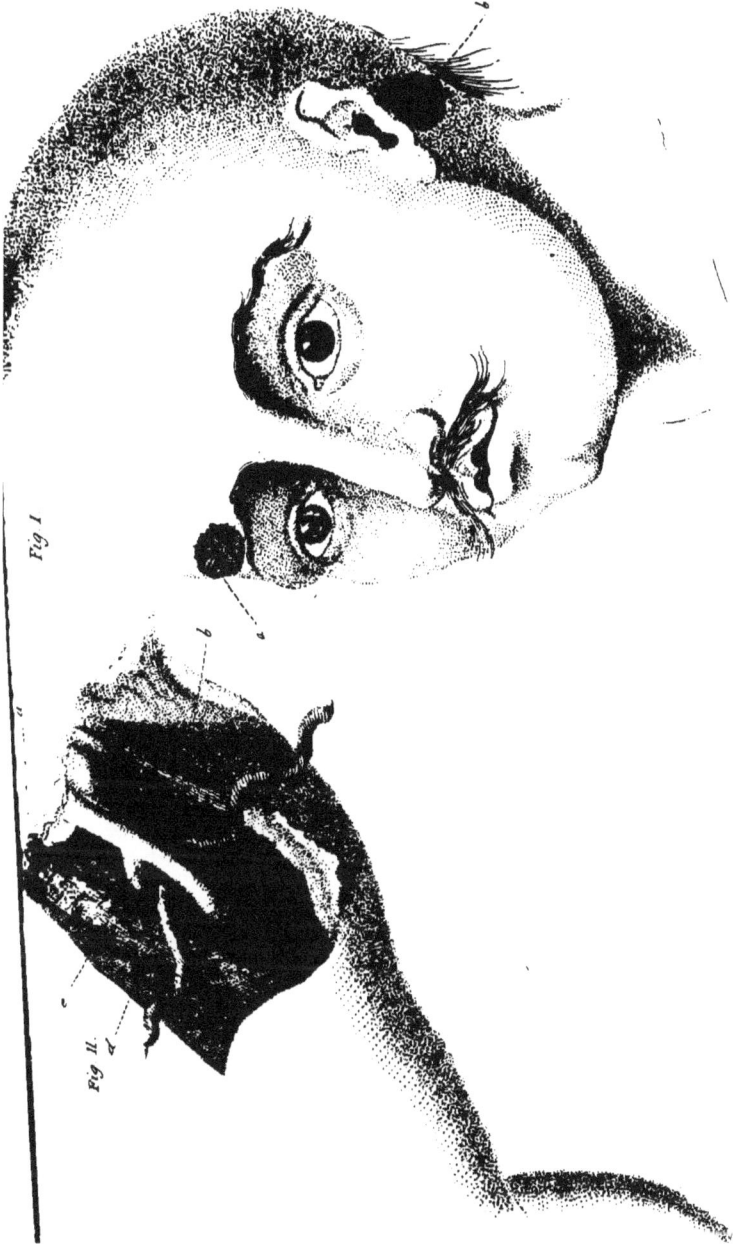

Fig I.

Fig II.

dreffings were applied; and his pulfe being full, he was let blood; after which he took twenty-five drops of tincture of opium. The next day he had a fenfe of heavinefs over his eyes, and obferved that objects did not appear to him fo brilliant as ufual; towards the evening he complained of naufea and thirft. He took *tart. vitriol.* and *antim. diaph.* $\bar{a}\bar{a}$ gr. xii every third hour, and a clyfter was adminiftered. On the third day he complained of pain of his head, accompanied with drowfinefs; and, at intervals, of a weaknefs of his extremities. As the clyfters had failed to procure a fufficient difcharge of fæces, he was directed to take three grains of calomel and fifteen grains of powder of jalap, which operated well, and procured an alleviation of the fymptoms juft now mentioned. His eyes were but flightly inflamed, and he complained of but little pain in that on the affected fide.

On the 6th day there was a good difcharge of matter from the wound, and efcars began to feparate in pretty large floughs. From this time he refted tolerably well without the ufe of the opiate, which till now had been repeated at bedtime. Splinters of bone, that had been driven in at the fuperior wound by the ball, came

away

away from the dependent orifice at almoſt every dreſſing (which was twice a day) for ſeveral days. The nauſea, head-ach, weakneſs of his limbs, thirſt, and every ſymptom of fever, gradually vaniſhed; the ſuperior orifice filled up with new granulations, and cicatrized firmly; and in about ten weeks there remained nothing more neceſſary than a ſuperficial dreſſing to the inferior opening near the ear.

I did not ſee this man after he had actually left off every application to the affected part; but from the condition of the wound, and the patient's health and vigour, I have not any room to doubt, that in a few days, after I laſt ſaw him, he was capable of returning to his duty.

On reflecting on this extraordinary injury, (inaſmuch as it was not a mortal one) I am inclined to think, that as the ball, though a large one, entered low down upon the orbit, and near the external part of the eye, it miſſed the os planum and frontal ſinuſes, and conſequently that branch of nerves that paſſes through them; ſo that, judging from its apparent direction, it muſt have injured part of the os ethmoides, near the ſeptum naſi. To this courſe of the
ball,

ball, and the favourable situation of the dependent orifice, the favourable event of the
cafe was probably owing; for though he complained at certain periods of a fenfe of weight
upon the upper and fore part of the head, general weaknefs of his limbs, and lofs of fight,
fymptoms indicating an oppreffion of the brain,
yet upon opening the wound, and giving vent to
the matter, which was in fome meafure confined
by the dreffings, thofe fymptoms gradually va-
nifhed, and the patient always became perfectly
eafy after the application, for a few minutes, of
a warm fomentation.

An inftance of a ball entering under the
right eye, and paffing obliquely through the
cerebrum and cranium above the right ear,
without hurting the eye or fight, is recorded
by Heifter in his Medical, Chirurgical, and
Anatomical Cafes and Obfervations, page 7
(of the Englifh tranflation) Obf. VII.

V. *An*

V. *An Account of some extraordinary Symptoms which were apparently connected with certain morbid Alterations about the Veins and Nerves. Communicated in a Letter to Dr.* Simmons *by Mr.* John Pearson, *Surgeon of the Lock Hospital, and of the Public Dispensary.*

MRS. P. aged fifty-one years, of Miles' Lane, Cannon Street, began to suffer from a peculiar uneasiness at the inner part of her left leg, about seventeen years ago, when she was in the third month of her second pregnancy. The skin which covered the particular seat of her complaint, retained its natural colour; but there was a circular induration, of about half an inch in diameter, very little elevated above the surface, which was exquisitely painful when slightly touched or compressed; this morbid part was situated in the course of the vena saphena major, and was about six inches above the joint of the ancle. Besides the acute pain which was produced by inadvertently touching this little tumour, Mrs. P. commonly suffered several paroxysms

of

of pain every day; each of thefe a'tacks was
accompanied with an increafed rednefs, and a
fenfible elevation of the indurated part, the
pain at the fame time extending to the knee,
and often darting to the ftomach ; the duration
of the fit was about twenty minutes ; it was attended with flight convulfive motions of different parts of the body, and frequently terminated with flatulent eructations. Thefe fits of
pain did not recur at any regular periods; fo
that the number which fhe underwent in the
courfe of a day was various and uncertain ; for
a difordered ftate of the ftomach, or a fudden
perturbation of mind would at any time excite
one of the paroxyfms. She alfo had obferved,
that the feverity of her fufferings was invariably
increafed during the periods of menftruation
and of pregnancy ; and that in the latter months
of geftation, the duration of each recurrence
of pain was extended to an hour and a half.
But although this difeafe was uniformly aggra-
vated by certain alterations in the ftate of the
uterus, yet it continued with undiminifhed feverity after Mrs. P. had ceafed to bear children; for when her youngeft child was more
than fix years old, fhe had not experienced any
abatement of her daily fufferings. About thir-

teen years ago, I advised her to have the morbid part removed; but at that time she was unwilling to undergo an operation; she however
submitted to various methods of treatment,
under the direction of different medical gen-
tlemen, but without obtaining any relief.

In the month of April, 1793, Dr. Lowder,
who had been long acquainted with the circum-
stances of this painful complaint, informed
Mrs. P. of the success which had attended the
removal of a similar tumour, by the application of a caustic. She read the history of the
case, which is published in the third volume
of the Memoirs of the Medical Society of London, and very soon determined to seek relief
from the same mode of treatment.

Accordingly, on the 22d of April, I applied
the lapis infernalis to the morbid part; she en-
dured the most excruciating tortures during several minutes after its application; but the pain
gradually diminished with the sensibility of the
part, so that in about twenty minutes the eschar
was completely formed, and she then felt no
more pain than what is the usual consequence
of a caustic applied to any part of the body.
From this day she never experienced the recurrence of a single paroxysm of pain; the eschar

2   exfoliated

exfoliated in about twelve days; and on the 7th of June the fore was perfectly healed.

As the preceding hiftory contains fome curious and rather uncommon circumfiances, I beg leave to offer a few obfervations upon fome of them. The indurated part having been deftroyed by a cauftic, it was not in my power to examine its internal ftructure, fo ·as to difcover the true nature of the morbid alteration. I afcertained, however, that a portion of the vena faphena major, and that branch of the crural nerve which accompanies it in its courfe down the infide of the leg, were completely included within this tumour. This fact was clearly demonftrated, after the exfoliation of the efchar; for I then faw a portion of the vein hanging down at the fuperior part of the fore, and the naked nerve in contact with it; and on touching the nerve with my probe, Mrs. P. inftantly complained of an acutely painful fenfation, fimilar to that which fhe had been accuftomed to feel before the tumour was removed. I then deftroyed that part of the nerve which was expofed with lunar cauftic, and my patient fuffered no more uneafinefs. After thus proving

H 2   that

that a vein, and a confiderable ramification of
a nerve, were contained within the difeafed
part, I proceed to obferve, that the paroxyfms
of pain were excited by every thing that acce-
lerated or otherwife difturbed the circulation of
the blood ; whether applied to the induration,
or affecting the general fyftem ; as all ftrong
exertions of the mufcles, external impulfe, or
mental commotion. The afcent of the blood,
in the veins of the lower extremities, is necef-
farily impeded in the ftate of pregnancy; and
during this period, the fits of pain were always
fharper, and were alfo of longer duration ; and
at the time of parturition, when the action of
the heart and blood-veffels. is confiderably increafed, Mrs. P. fuffered exceedingly; for, to
ufe her own expreffion, fhe " had all her labour
pains in her leg."

It is alfo highly probable, that the portion
of vein which paffed through the tumour was
unufually diftended with blood at the time of
the paroxyfm ; for upon thefe occafions, the
morbid furface became redder than common;
and the tumour was fenfibly elevated. We may
therefore, perhaps, venture to conclude, that the
vein and the nerve being confined within a fubftance that could not be eafily diftended, when-

ever

ever the vein became preternaturally turgid, the nerve was compreſſed between its parietes and the internal ſurface of the induration; and that conſequently the ſymptoms were connected with this ſtate of the part. I do not ſuppoſe that it will be neceſſary for me to undertake a proof in detail, that a certain degree of preſſure upon a nerve will produce pain, ſpaſms, and convulſions; it may be ſufficient for my purpoſe to refer to a few of the many inſtances which are recorded in medical books. In the fourth volume of the Edinburgh Medical Eſſays, Dr. Short has related the hiſtory of an epilepſy, which was cauſed by the preſſure of a hard cartilaginous ſubſtance upon a nerve; he cured his patient by removing the tumour, and dividing the nerve. Guattani, in his Treatiſe *de externis Aneuryſmatibus,* (Hiſt. XX.) has recorded a caſe in which violent ſpaſms were occaſioned by the preſſure of an aneuriſm upon a nerve. In the Eſſays and Obſervations Phyſical and Literary, Vol. III., the late Sir John Pringle has publiſhed a Caſe, where a tumour formed by extravaſated blood, by preſſing upon the intercoſtal nerves, produced pain, irritation, and perhaps a hic-

H 3 cup,

cup, which could not be ſtopped*. I do not intend to deduce any general concluſion from a particular inſtance; for although the remarkable ſymptoms which occurred in Mrs. P.'s caſe, were connected with a morbid ſtate of a vein and a nerve; yet as no account has been publiſhed of the internal ſtructure of parts which have been affected by a ſimilar complaint, it would be improper to conclude, that every inſtance of local morbid ſenſibility, accompanied with convulſive motions and pain, muſt depend upon ſuch a peculiar condition of the ſuffering parts. I have indeed ſeen another caſe, very much reſembling that of Mrs. P.'s, in which there is a ſmall exquiſitely ſenſible induration

---

* For inſtances of convulſive motions, and even epilepſy, produced by local diſeaſes about ſome of the extremities, or that were cured by the removal of matter, carious bone, or ſome portion of the integuments, conſult Willis *de Morlis Convulſ.*; Riverius *de Epilepſia*; Schenckii O*ſervat.* (Lib. *de Epilepſia.*); Foreſtus *de Cerebri Morbis*, Lib. X. Obſ. 67; Petri Borelli *Hiſtor. & Obſervat. medico phyſicarum*, Cent. II. Obſ. 95. Joh. Rhodii *Obſerv. Med.* Cent. I.; Tulpii *Obſerv. Med.* Lib. IV. Cap. 2; Boneti *Sepulchretum*, Lib. I. Sect. 13; Van Swieten *Comment. in Aph.* H. Boerhaave, Tom. III. § 1075. Haller *Elementa Phyſiologiæ*, Tom. IV. § 30. Simſon on the Vital and Animal Actions, Eſſay I. ch. 3.

at

at the pofterior part of the leg, near the beginning of the tendo achillis, from which the patient fuffers acutely whenever it is touched. She
has occafional paroxyfms of pain, but they return at uncertain intervals; and fhe thinks that
they grow milder. In this inftance, as in that
recorded by Dr. Biffet *, the tumour becomes
uneafy in rainy and windy weather; but it does
not appear that the difeafe had ever any con-
nexion with pregnancy. I fufpect that the
tumour, which I have juft now mentioned, may
be connected with the vena faphena minor,
and that confequently it may include or comprefs a fmall branch of the fciatic nerve; but
as I could not render the cutaneous veins of
the leg turgid by moderate preffure, its exact
fituation was not afcertained †.

In

* Memoirs of the Medical Society of London, Vol. III.
Art. VI.

† The firft volume of M. Pouteau's pofthumous works
contains a very curious hiftory of a difeafe which he there
calls cancerous; whether properly or no I fhall not inquire;
but as it refembles Mrs. P.'s cafe in fome of its characters,
I fhall take the liberty of prefenting an abftract of it:

" On voyoit à la partie baffe du *Sternum* une furface
" ovale de largeur d'un ecu de fix livres dans fon petit dia-

H 4 " metre,

In the early part of the laſt Spring, a young
married woman applied to me at the Public Diſ-
penſary, complaining of pain and lameneſs of the
right arm.   She ſhewed me a tumour of a pale
red colour, and of about the ſize of a filberd,

" metre, ſans elévation, ſans rougeur, ſans engorgement*
" circonvoiſin.   La peau ſeulement qui la recouvroit étoit
" un peu moins nette, que par tout ailleurs, mais ſemblable
" à la ſenſitive qui paroit craindre la main qui l'approche.
" Cette portion des tégumens auroit fait reſſentir les plus
" vives douleurs, ſi le doigt, ſans la toucher, en eût ap-
" proché avec trop de célérité.   Le moindre inſecte, un
" fetu que le haſard auroit fait poſer deſſus, euſſent auſſi-
" tôt rappelié les convulſions.   Les rétours de ces convul-
" ſions étoient periodiques, ſe montrant à ſept heures &
" demie préciſes du ſoir.   Dans le plus grand calme, on ne
" les attendoit que de deux jours l'un ; & à la moindre agita-
" tion, les mouvemens convulſifs étoient journaliers. Leur
" durée étoit de deux heures, & même plus."   The hiſtory preſents us with many other extraordinary circu-
ſtances; but it may be ſufficient at this time to add, that
M. Pouteau made a crucial inciſion in this morbidly ſenſible part, which afforded an immediate although but a temporary ſuſpenſion of the pain and convulſions.   He then
extirpated the portion of diſeaſed integuments ; but as the
young lady was not perfectly relieved by this operation, he
finally completed the cure by burning a cylinder of cotton
upon the part.   Vide *Oeuvres poſthumes de M. Pouteau*,
Tom. I. ch. I.

which

which was fituated in the courfe of the vena
mediana bafilica, at the bend of the arm : this
morbid part was conftantly uneafy; but when
it was preffed or handled, fhe complained of
acute pain, which extended along the upper
arm, and produced flight convulfive motions
in the mufcles. She derived no advantage from
mild difcutient and emollient applications ;
but her pain increafed fo much, that her health
became injured, and fhe was at length confined
to her bed. On vifiting her at home, I found
the tumour unaltered in its appearance, excepting a fpontaneous feparation of the cuticle from
its furface ; fhe was in conftant pain ; the uneafinefs not only proceeding along the upper arm,
but alfo to the neck, and affecting the breaft and
mufcles on the right fide. Her pulfe was feeble,
but not too frequent ; fhe complained of a great
fenfe of weaknefs, and convulfive motions were
excited in the mufcles of the upper arm, neck,
and thorax, on that fide, by the gentleft examination of the morbid part. I ordered a large
veficatory to be applied on the inner part of
the fore arm, and directed her to take ten
grains of *pulvis ipecacuanhæ compofitus*, whenever her pain fhould be unufually fevere. She
foon derived confiderable relief from this mode

of

of treatment : the bliftering plafter was repeated twice during my attendance; the tumour gradually became lefs painful, and diminifhed in bulk ; and in about a month it had entirely difappeared. It was not more than three weeks after fhe was difmiffed, when fhe applied to me again, on account of a tumour very much refembling the former one, which was fituated at the bend of the arm, in the courfe of the vena cephalica; fo that a portion of the vein evidently paffed through, or, rather, was included within the center of the morbid part. The pain and morbid irritability affected the fame parts as before, but in a much inferior degree. I directed a mode of treatment fimilar to that which had been employed on the former occafion, and it was attended with equal fuccefs.

This young woman had fome fymptoms which indicated a difeafed ftate of the lungs; and fhe occafionally fpat blood : but fhe had not been formerly fubject to any particular complaints; fhe menftruated regularly ; and had never been pregnant. I cannot affign any probable caufe for the appearance of fo fingular a complaint as that which I have now defcribed ; but fome of the effects which took place would

perhaps

perhaps admit of an explanation, if it could be
proved that a fmall ramification of a nerve, as
well as a portion of a vein, were included within
each of the tumours. That this was actually the
cafe is highly probable, becaufe the cutaneous
nerve diftributes feveral of its branches in the
vicinity of the vena mediana bafilica; and fmall
fibrils belonging to the mufculo-cutaneous nerve,
are commonly feen near the vena cephalica, and
the vena mediana cephalica; fo that tumours
fituated at the bend of the arm, and in the
courfe of thefe blood veffels, muft be almoft ne-
ceffarily in contact with one or more branches
belonging to the internal, or external cutaneous
nerves *.

* The late Profeffor Camper, in a valuable work
entitled *Demonftrationum Anatomico-Pathologicarum Liber
primus continens Brachii humani Fabricam et Morbos*, has
given a very diftinct view of the mode in which thefe
fmall branches of nerves are diftributed at the bend of the
arm; and his engravings are accompanied with fome good
practical obfervations. Mr. Abernethy alfo publifhed two
engravings, laft year, in the fecond part of his Surgical and
Phyfiological Effays, in which the courfe of thefe nerves
is very neatly and correctly delineated: and the effay to
which they are annexed, contains many ufeful remarks
" on the ill confequences fometimes fucceeding to venæ-
" fection."

I beg

I beg leave to refer it to the intelligent reader, how far the following account of a difeafe of the fubcutaneous nerves, as defcribed by Profeffor Camper in the work already referred to, bears any refemblance to the preceding hiftories.

" Non raro in nervis cutaneis tubercula par-
" va ac dura obfervantur, quæ vera ganglia
" funt, pifi magnitudinem licet non excedant;
" dies tamen noctefque acutiffimis lancinanti-
" bus doloribus ægros torquent : externis re-
" mediis non cedunt ; fcalpello igitur ea attin-
" gere oportet. Francqueræ ex cubito feminæ
" tale, plagâ factâ, fuftuli, quod ramo mufculo-
" cutanei nervi adhærebat : poft operationem
" optime fe habuit. In fubcutaneis nervis fre-
" quenter effe videntur. Amftelædami fimile
" ganglium genu mulieris occupans, eodem
" modo fanari curavi. In viris plus femel ea
" vidi : albicant intus, cartilagineæ duritiæ
" funt, renitentia, & intra nervorum tunicas
" fedem habent." Lib. I. P. 11. Cap. 2. § 5.

I have feen many fymptoms refembling thofe which occurred in the preceding cafes, apparently follow, as confequences of wounds inflicted on fmall branches of nerves ; but as this paper is already much longer than I expected it would have been, I muft defer giving an account

count of them to another opportunity. As the
following cafe exhibits fome uncommon circumftances, I infert it as a kind of fupplement to
the foregoing hiftories.

"The fingular effects of an iffue in the in-
"fide of the thigh, which appeared in the cafe
"of a clergyman; written by himfelf, Auguft
"25th, 1793.

"The Rev. Dr. T——, of Knightfbridge,
"above 60 years of age, having had a hint
"from a medical friend, that an iffue might be
"of ufe to his health, he had one made by a
"blifter, in the lower part and at the infide
"of his right thigh, about the end of May laft.
"Two days after the pea was put in, he was
"feized with a ficknefs and vomiting, which
"continued feveral hours. In about fix days
"after this firft attack, he had a return of the
"fame fymptoms; and thefe fits recurred every
"fix or feven days. But what is very remark-
"able, when the iffue began to difcharge, he
"became deaf in both his ears, and the deaf-
"nefs arrived to fuch a degree, that in preach-
"ing he could but juft hear his own voice.

"After the iffue had been kept open fix
"weeks, it occurred to him, that perhaps the
"regular fits of ficknefs and vomiting, and the
                                    "unufual

" unufual deafnefs, (both of which he recol-
" lected had commenced with the iffue) were
" occafioned by a fympathy of the nerves ; and
" having made obfervations for one week longer,
" which confirmed this opinion, he determined
" to dry it up. This he did gradually, by ufing
" peafe of a fmaller fize, till the ulcer was not
" more than one eighth of an inch in diameter.
" When the pea had been out only twelve hours,
" he was fenfible of fome fmall return of his
" hearing, and on looking at the fore, he found
" it healed ; which he confidered as a farther
" confirmation of his opinion, refpecting the
" caufe of his deafnefs, as well as of the ficknefs
" and vomiting. He found, that as the wound
" healed, the deafnefs leffened, and when it was
" completely healed, his hearing was quite reco-
" vered, nor has he had one fit of ficknefs fince."

When Dr. T—— related his cafe to me, I
defired him to let me fee the cicatrix of the
iffue ; and on carefully examining it, it appeared probable that the pea had preffed againft
the fide of the vena faphena. I would alfo farther add, that my examination of the part excited a flight degree of naufea.

VI. *An*

VI. *An Account of the Extraction of an extraneous Substance from the Rectum. By Mr.* William Blair, *Surgeon of the Lock Hospital; and of the General Dispensary in Newman Street, St. Mary-le-bone.*

ON Tuesday, the 25th of March last, a French gentleman was sent to me by an Apothecary in this neighbourhood, complaining of a pungent, hot, and irritating sensation in the *rectum*; which was considerably augmented during every evacuation *per anum*. These painful symptoms had commenced on the preceding Sunday, and continued to encrease in so alarming a manner, that, upon the day following, he was induced to examine with his finger, whether or not any foreign substance, or other cause of his uneasiness, could be discovered in the intestine. He had the good fortune to feel something in the rectum, which he thought was unnatural, but could not remove it; and therefore he applied the next day for chirurgical assistance.

Having submitted the patient to a proper examination, I readily perceived an hard body confined in the interior membrane of the intestine. With the help of a pair of forceps, I
extracted

extracted two portions of a brittle black fubftance; which, on careful infpection, appeared
to be bread toafted nearly to a cinder : the two
pieces, which were whole before the extraction
was attempted, might be together about an inch
in length, half an inch in width, and one third
of an inch in diameter.

The patient remembered to have fwallowed
fomething with confiderable difficulty two
days before, while partaking of fome foup;
which was probably the fame morfel of bread
that diftreffed him upon this occafion.

Does it not appear from this cafe, that bread
when toafted is lefs fit for digeftion than fome perfons would have us believe ; and that it affords
but little nourifhment compared with that which
is moderately baked ?

However trifling the circumftances of the
above cafe may be regarded in its earlieft ftage;
there can be no doubt entertained of the probability of its terminating very ferioufly, if the
patient had not applied for fpeedy relief : inflammation, abfcefs, and all their confequences,
might have enfued, if the efforts of nature, or
the power of aperient and antiphlogiftic remedies had alone been trufted to.

In fimilar inftances, without lofing time by
endeavours

endeavours to relieve the patient's fufferings by
medicine, it will be immediately proper to fubject him to a careful examination. If the fimple
introduction of a finger be infufficient to difen-
gage the extraneous body, and it can be felt
adhering to the *rugæ*, or piercing the coats of
the rectum, a pair of blunt-pointed fciffars, or
forceps, (as the cafe may indicate) fhould be
gently conducted upon the finger, in order to
divide, break in pieces, or loofen the foreign
fubftance : if a pointed bone, or other hard
and fharp body, fhould be confined acrofs
the gut, endangering the neighbouring parts,
it will be prudent to empty the urinary bladder, previous to any attempt to remove it by
mechanical means: and, fhould the pain, and
other ill effects become urgent, it might be neceffary, after milder methods had proved inef-
fectual, to make a judicious incifion either into
the rectum, or circumjacent integuments, as
the peculiarities of the cafe fhould require to fa-
cilitate the extraction. To obviate the inflammation, and its concomitant fymptoms, leeches,
anodyne and laxative clyfters, with the ufual
antiphlogiftic remedies, ought to be diligently
employed.

Inftances of the kind above related, with

VOL. VI.   I   fuitable

fuitable remarks, are recorded by feveral practical authors; but the reader may fpare himfelf
the trouble of perufing fome of them, by con-
fulting the *Memoires de l'Académie Royale de
Chirurgie*, Tom. I. p. 540, et feq. 4to Edit.

Newman Street, Oct. 6, 1794.

———

VII. *A Cafe of Aneurifm of the Crural Artery;
communicated in a Letter to Dr.* Simmons, *by
Mr.* Thompfon Forfter, *Surgeon on the Staff
of the Army, and Surgeon to Guy's Hofpital.*

TO DR. SIMMONS.

Dear Sir,

TO the two cafes of Aneurifm which you
have done me the honour to infert in
the fifth volume of Medical Facts and Obfervations, I am defirous of adding the following,
as I flatter myfelf it will tend ftill further to
elucidate the peculiar utility and advantages of
the operation in queftion.

Believe me, Dear Sir,

Your's, &c.

Nov. 3, 1794.   THOMPSON FORSTER.

CASE.

CASE.

Lawrence M'Carthy, a labouring man, aged thirty-feven years, was admitted, as my patient, into Guy's Hofpital, on the 30th of July 1794, for the cure of an aneurifm of the crural artery.

About nine months before his admiffion, he had perceived a fmall tumor on his right thigh, near that part where the crural artery dips under the triceps mufcle; as it occafioned no inconvenience, nor prevented his working, he took but little notice of it; it came fpontaneoufly, without any external violence, and remained ftationary for near fix months before it became painful: when the tumor had acquired the fize of an egg, a pulfation was perceptible in it, but not before.

At this period of the difeafe he was advifed to foment the part, and to make ufe of liniments: this he continued to do for fome time; but finding no relief from thefe remedies, he applied to a furgeon, who recommended the ufe of a bandage, which he made ufe of for near three months, but without any abatement of the pain; and the tumor in the mean time had increafed to a very confiderable

fize,

fize, and the limb in general had acquired fomething more than its natural bulk.

The patient, naturally hypochondriacal, became anxious, irritable, and dejected; complaining of great pain in the limb, and particularly in the tumor, which was in fome meafure eafed by preffure. In this ftate he came into the hofpital; and his general habit having been lowered by bleeding, purgatives, and a fuitable regimen previoufly to the operation, I performed it on Monday, the 11th of Auguft, by making an incifion in the courfe of the lower edge of the fartorius mufcle, and about an inch below where the profunda is ufually given off. Having laid bare the artery, * I paffed a ligature under it with an eyed probe, and applying the ftick, furrounded by adhefive plafter, &c. as defcribed in the former cafes †, the ar-

---

* With a view of conveying to the reader a more precife idea of the operation, I have made a fketch of the parts concerned in it, from a fubject diffected for the purpofe. See the annexed engraving (plate 1, fig. 2.) in which *a* refers to Poupart's ligament; *b* to the crural artery, with a ligature paffed under it at the part where it was tied; *c* to the profunda; and *d* to the fartorius mufcle. It feems hardly neceffary to remind the reader that the object of this fketch being merely to point out the feat of the operation, the parts are delineated in their natural ftate.

† Vide Vol. V. p. 6.

tery

tery was thus furrounded, and by thefe means equally compreffed; the pulfation below of courfe ceafed : but, for fear of a fudden hæ-morrhage, I paffed a fecond ligature about half an inch above the former, laying it loofe, that an affiftant might inftantly tie it in cafe of fuch an accident.

Auguft 21ft. The firft ligature, with the ftick, came away with eafe.

Auguft 22d. The fecond ligature came away with equal eafe.

An account of the ftate of the pulfe at the wrift, and of the temperature of both limbs, at the ham, and at the foot, was taken every day with great accuracy by Mr. G. Babington, according to the annexed Table *, until Auguft the 27th, when the temperature of each was found to be equal.

The fize of the tumor gradually decreafed, and the patient, having the perfect ufe of his limb, was difmiffed, cured, October 10, 1794.

The preceding cafe differs materially from the two former, not only in the circumftance of the tumor in this having been fituated in the

* See page 119.

I 3

upper

upper part of the thigh, fo that the artery could
not be fecured lower than about an inch below
where the profunda is ufually given off, but
likewife in the very great pain the patient en-
dured both night and day for three weeks before
the operation. The tumor was as confiderable,
but the enlargement of the limb below it was
much lefs than in the former cafes. After the
operation, the fymptoms were much flighter
than in the other cafes, probably owing to the
low ftate I thought it proper to reduce the patient
to for the purpofe; and the ligature came away
on the tenth day after the operation without the
leaft trouble. But the circumftance in which it
differed the moft effentially from the other two,
was, that the tumor was completely abforbed in
feven weeks, and the patient had then acquired
a perfect ufe of the limb, while, in the former
cafes, the patients did indeed acquire the ufe of
their limbs, but the tumors, though leffened
and free from pulfation, ftill remained.

TABLE.

TABLE.

Day of the Month.	pulse at wrist	tem. of arm.	tem. right ham.	right foot.	left ham.	left foot.	Time of day when the obs. were made.
August 11		68½°	98°	94°	97°	96°	10½ P. M.
12		68½	97	91	91	89	8½ A. M.
	128	70	99	91	94	93	10½ P. M.
13	109	68	98	92	90	88	8½ A. M.
	112	71	100	95	98	95	10½ P. M.
14	104	68	98	91	91	91½	8½ A. M.
	116	72	99	96	96	96	8½ P. M.
15	96	69	97	91	94	88	8½ A. M.
	112	72½	97	93	94	95	8 P. M.
16	97	72	98	93½	94	90	8 A. M.
	112	73	98	95	94	94	8 P. M.
17	96	71	98	92	95	89	8 A. M.
	112	74	97	94	94	94	8 P. M.
18	92	70	93	91	92½	89	8 A. M.
	110	72	97	92	94	93	9 P. M.
19	100	68	94	90	92	91	8 A. M.
	124	71½	101	96½	97	97	8½ P. M.
20	114	67	100	93	96	94	8½ A. M.
	116	70	99	96	95	94	8½ P. M.
21 First ligature and stick came away with ease, there being a perfect solution of continuity.							
	100	66°	97°	8°	93°	86°	8 A. M.
	100	69	98	92	95	94	8 P. M.
22 Second ligature was removed.							
	100	69°	96°	86°	93°	84°	8½ A. M.
	108	69	98	94	97	95	9 P. M.
23	100	67½	96	91	93	90	9 A. M.
	104	70½	98	94	95	93	8½ P. M.
24	96	69	97	89	95	87	10 A. M.
	104	69	99	95	98	95	8 P. M.
25	104	66½	98	93	96	92	8½ A. M.
	106	69½	95	92	93	91	8 P. M.
26	100	64	96	90	91	86	8 A. M.
	106	66	98	92	94	90	8½ P. M.
27	100	64	96	91	92	91	8½ A. M.
	95	63½	96	90	96	90	8 P. M.

VIII. *An Account of a Key Instrument of a new Construction ; with Observations on the Principles on which it acts, in the Extraction of Teeth, and on the Mode of applying it. By Mr.* Robert Clarke, *Surgeon at Sunderland, in the County of Durham. Communicated in a Letter to Mr.* Anthony Carlisle, *Surgeon of the Westminster Hospital, and Reader of Anatomy in London; and by him to Dr.* Simmons.

To Mr. CARLISLE.

SIR,

WITH this I send you a Key Instrument, for the Extraction of Teeth, of a construction different from any in common use, and which in practice fully answers to the expectations I had formed, *a priori*, from a careful examination of the principles of its action.

I cannot, perhaps, give you a clearer idea of its advantages, than that which you will obtain by pursuing the same train of investigation which I followed myself. I shall therefore proceed to lay it before you, that I may more thoroughly convince you of the propriety of the alteration I have made, or be corrected by your pointing out any error I may have fallen into.

In

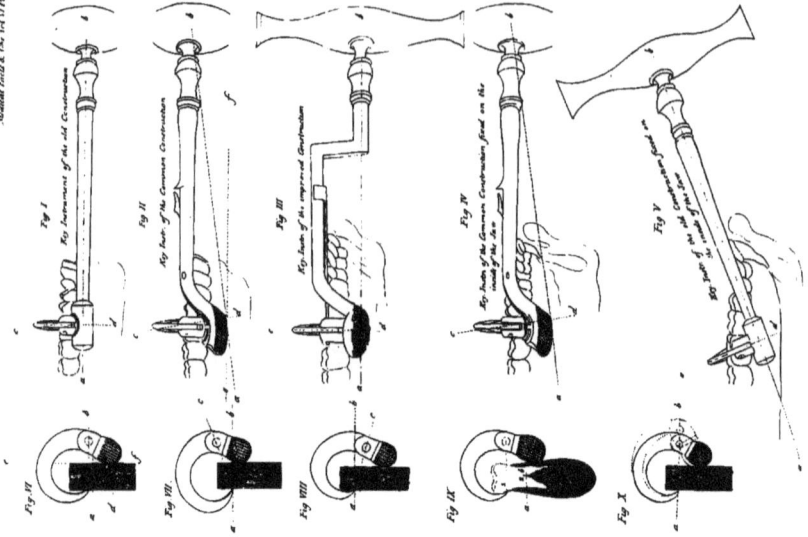

Fig. I
Key Instrument of the old Construction

Fig. II
Key bent of the Common Construction

Fig. III
Key bent of the improved Construction

Fig. IV
Key bent of the Common Construction fixed on the inside of the Jaw

Fig. V
Key bent of the old Construction fixed on the inside of the Jaw

Fig. VI

Fig. VII

Fig. VIII

Fig. IX

Fig. X

In the firſt place then, it appeared to me that as the fulcrum, or point, upon which the tooth is carried round as on a center, is that part of the bolſter which reſts upon the gums, the axis of motion of the inſtrument would always be found by drawing a line through that point and the middle of the handle; and confequently that the old conſtruction of the Key Inſtrument was free from an inconvenience which attends the more modern one : I mean the axis of the bolſter and axis of the ſhank making an angle with each other ; on which account it is diſpoſed to ſhift its point of action on the gums, and to raiſe the tooth in a plane inclined to the throat, inſtead of a vertical one, as may be clearly ſeen by infpecting Figures I. II. (Plate II.*) where *a*, *b*, repreſent the axis of motion; *c*, *d*, the direction in which each inſtrument raiſes the tooth ; and *e. f.* (Fig. II.) the axis of the bolſter.

Now as the line of direction in Fig. I. is perpendicular to the jaw, it is needlefs to ſay that it is highly preferable to Fig. II. where the line of direction is inclined backward, making the

* It feems right to obferve here, that all the figures of this plate are on a reduced fcale of two thirds of their proper fize.

extraction

extraction of the tooth more difficult, and expofing that which is fituated behind it to be
driven from its focket, or even to be caught in
the arch of the claw. Befides this, the bolfter
refts only upon the corner *d*, adding greatly to
the injury of the gums.

The conftruction then of the Key-inftrument
delineated in Fig. I. would feem perfect, were it
not that in drawing teeth inwards, with refpect
to the jaw, the fore teeth prevent its due application, confining it to the direction reprefented
in Figure V.

To remedy this imperfection I have made
the inftrument with a bend in its fhank, to clear
the fore teeth, and to allow its proper application,
as in Figure III. where the fame obfervations
and references apply as in Figure I. and therefore it is unneceffary to repeat them. But in
order that the comparative merits of the three
inftruments may be feen at a glance, I have
added Figures IV. and V. wherein the axis of
motion, and the direction of the rifing tooth, are
fhown by dotted lines.

Having fully confidered what relates to the
direction of the tooth, I fhall next examine the
mechanifm which takes hold of it. For this
purpofe recourfe muft be had to the engraving.

Let

Let *a*, *b*, *c*, Figure VI. reprefent an end view of a Key inftrument, fixed upon a piece of hard, fmooth wood. Then it is obvious, that if it be turned from left to right, by means of its handle, it will break the wood in the direction *d*, *c*, and caufe the upper fragment to revolve on the point *c*, as a center. It is equally obvious, that if a line be drawn from the point *a*, croffing the oppofite furface of the folid *e*, *f*, at right angles, the counterpoife of the claw will fall into that line before it can take hold; for then the point *b*, is at the greateft poffible diftance from the furface *e*, *f*; confequently if the inftrument be placed as in Figure VII. the point *c* will defcend; or, if as in Figure VIII. it will afcend until it coincides with the line *a*, *b*.

I fhall now endeavour to apply this to practice. Let 1, 2, 3, in Figure IX. reprefent a tooth with its roots fixed in a fection of the jaw, and its corona engaged in a Key-inftrument; then it will readily appear that upon the action of the inftrument, the tooth will be drawn from its focket, and carried round the point *b*, as a center, rather than the joint fubftance of the tooth and jaw be broken in the line *a*, *b*, as happens in Figure VI. This however happens

3

only under particular circumſtances : For if the bolſter be placed too high, the tooth will be broken ; and if too low, the alveolar proceſs will always be torn away with it. It is therefore a matter of importance to determine the beſt point of contact for the bolſter, and this I have uniformly found to be at two-thirds the depth of the tooth, the claw being fixed at one third, as repreſented in Figure IX.

It will always be eaſy to aſcertain this point, by attending to the ſize of the corona, and the part of the jaw where the tooth is ſituated ; and equally ſo to make the inſtrument act upon it, by uſing a larger or ſmaller claw as the caſe may require. For illuſtration, however, I ſhall refer to Figure X. which repreſents a piece of wood graſped by the tooth inſtrument in the ſame manner as in Figure VI. Now if a larger claw, ſhewn by the dotted line, be uſed, the bolſter will fix higher upon the wood than before. For as the center pin of the claw will always reſt in the line a, b, the bolſter muſt riſe higher before it can come into contact. But notwithſtanding the uſe of a larger or ſmaller claw, in proportion to the ſize of the tooth, enables us to fix it at a proper height, the uſe of a very diſproportionate one is always

inconvenient,

inconvenient, by depriving us of the ufe of the crank, in drawing teeth inwards, and by encroaching upon the cheeks in drawing them outwards. I have therefore in the conftruction of this inftrument, taken care to make the bolfter of fuch a depth, as to be free from either inconvenience.

The form of the bolfter is by no means a matter of indifference; for if it be too fmall, it prefents fo fmall a furface to the gums, that the preffure made upon them, by the extraction of a tooth moderately firm, cuts them through, and even penetrates the bone, efpecially if the bolfter be of the ufual form. I have therefore been careful to make it of a proper fize, and to give it a prolate fpheroidal figure, as being the leaft difpofed to injure the gums, and applicable with exactnefs and eafe to all parts of the mouth; and in order ftill further to guard againft this bruifing of the gums, I wrap the bolfter to the thicknefs of a line, with tow, wound on as tight as I can, before I flide forward the bolt and put in the claw.

I have alfo been attentive to the form of the claws, that they may touch the tooth only with their points. And the inftrument is fo contrived, that they can be quickly changed or

turned

turned to an oppofite direction as the cafe may require : this is done by means of a fliding bolt, inftead of a fcrew, which paffes through the claws.

I have always found that when the tooth is to be turned from right to left in drawing it, that the handle anfwers beft placed perpendicularly; and when from left to right, horizontally. The reafon of this will be obvious, if we confider that in the firft cafe, the pronator mufcles of the operator's arm, which are thofe exerting the force, act with moft advantage when the hand is vertical; and in the fecond cafe, that the fupinators act moft advantageoufly with the hand prone. I have therefore contrived the handle fo that it may be eafily turned, as often as there is occafion to turn the claw.

I am, Sir, &c.

Sunderland,   Robert Clarke.
Aug. 18, 1794.

IX. *An*

IX. *An Account of a new Species of Swietenia (Mahogany); and of Experiments and Observations on its Bark, made with a View to ascertain its Powers, and to compare them with those of Peruvian Bark, for which it is proposed as a Substitute: Being an Abstract of a Paper on this Subject, addressed to the Honourable Court of Directors of the United East-India Company.* By William Roxburgh, *M.D.*

THE species of Swietenia described in this paper, and which Dr. Roxburgh names *Swietenia Febrifuga* *, is a native of the mountainous part of the Rajamundry Circar, North of Samulcotah and Peddapore. It is a very

---

* Dr. Andrew Duncan, junior, who has made this new species of Swietenia the subject of a very ingenious inaugural Differtation, gives a good reason for preferring, as a trivial name, the Hindoo appellation, *Soymida*, to one founded on its medicinal properties; similar properties, he observes, being ascribed by Dr. Wright (London Medical Journal, Vol. VIII. p, 286) to the mahogany tree of Jamaica *(Swietenia Mahagoni)*, another species of the same genus.—Vide *Tentamen inaugurale de Swietenia Soymida* ; *Auctore* Andrea Duncan. 8vo. Edinburgi, 1794. EDITOR.

large

large tree, known among the Hindoos by the name of *Soymida*, and flowers about the end of the cold or beginning of the hot feafon. Its feeds ripen in three or four months after.

Of this tree Dr. Roxburgh gives the following botanic defcription :

"TRUNK. Very ftraight, rifing to a great
"height, of a great thicknefs, and covered
"with a grey, fcabrous, cracked bark.

"BRANCHES. Numerous, the lower
"fpreading, the higher afcending, forming a
"very large fhady head.

"LEAVES. Alternate, about the extre-
"mities of the brachlets *(turiones)* abruptly
"feathered, about twelve inches long.

"LEAFLETS. Oppofite, very fhort, pe-
"tiolated, three or four pair, oval, obtufe, or
"end-nicked, the lower fide generally extend-
"ing a little further down on the petiolet than
"the upper; fmooth, fhining; from three to
"five inches long, and from two to three
"broad, the inferior fmalleft.

"PETIOLE. Round, fmooth, about nine
"to ten inches long.

"STIPULES none.

"PANICLE. Very large, terminal, dif-
fufe,

" fufe, bearing great numbers of middle-fized,
" white, inodorous flowers.

" PEDUNCLE and PEDICLES. Round
" and fmooth.

" BRACTS. Very minute.

" CALYX. Below, five-leaved; LEAF-
" LETS. Oval, deciduous.

" COROL. Petals five, inverfe, egged,
" obtufe, concave, expanding. NECTARY.
" Not quite half the length of the petals, a
" little bellied; mouth ten-toothed, teeth bi-
" fid (two-cleft).

" STAMEN. Filaments ten, very fhort,
" inferted juft within the mouth of the nectary.
" ANTHERS. Oval.

" PISTIL. Germ conical. STYLE. Thick,
" tapering. STIGMA. Large, targetted, fhut-
" ting up the mouth of the nectary.

" PERICARP. Capfule egged, large, five-
" celled, five-valved, valvelets gaping from
" the top.

" RECEPTACLE. In the centre, large,
" fpongy, five-angled; angles fharp and con-
" nected, with the futures of the capfule, be-
" tween them, deeply fulcated.

" SEEDS. Many in each cell, imbricated,
" obliquely wedge-fhaped, enlarged by a long

" membranaceous wing, inferted, at the upper
" point of the wing, into a long brown fpeck
" on the upper part of the excavations of the
" receptacle : all the reft of the receptacle is
" white."

The wood of this tree, we are told, is of a
dull red colour, remarkably hard and heavy;
and is reckoned, by the natives, by far the
moft durable timber they know; on which account it is ufed for all the wood work in their
temples.

The bark of the trunk and large branches,
of large and middle-fized trees, is covered
with a dark rufty-coloured coat, of about an
eighth of an inch in thicknefs, which cracks in
various directions, and fometimes peels off in
irregular pieces, according to the directions of
the cracks. Immediately under this is a very
firm, but brittle coat, of about three-eighths
of an inch in thicknefs. When firft cut, it is
light-coloured; but on drying, or even expofure to the air for a few minutes, it turns to a
reddifh brown. The inner lamina are thin,
confifting of tough, lighter-coloured layers.

The bark of the younger branches is not
cracked, is pretty fmooth, of a much lighter
colour,

colour, and has not the rufty coat above defcribed, but has often many blotches of various
coloured lichen over it: it confifts wholly of
the brown, folid, and inner layers.

The outer ruft-coloured layer of the trunk
has but little tafte; the other two poffefs a little aromatic fmell, and their tafte is very bitter
and aftringent, accompanied with fomething
aromatic, but in a trifling degree. There is
nothing difagreeable in the tafte, more than
may be expected from a pure, fimple, ftrong
bitter and aftringent united. The middle lamina are eafily reduced to a very fine rofe or
light brown-coloured powder.

Cold water, in the courfe of an hour, our author obferves, acquired from this bark a deep but
clear reddifh colour. The moft minute portion of
a chalybeate (one drop of a folution of twenty
grains of fal martis in an ounce of water) inftantly changed a much-diluted cold infufion
to a deep purple, which, on ftanding, became
darker and darker, with a reddifh tinge; and
no decompofition took place till about the tenth
day; the colouring matter then began to feparate, and fall to the bottom in black flakes,
leaving the liquor almoft colourlefs. If the
infufion was fome days (from four to thirty)

old,

old, the colour produced by the martial folution was as inftantaneous as when frefh, and deeper; and at no peiod, up to thirty days, did it fhow the leaft tinge of green. Ten times the fame quantity of the fame martial folution, it feems, did not produce fo great a change upon a fimilar infufion of the common pale Peruvian bark; and its effect on the latter was much flower. Its bitter qualities are alfo defcribed as much more intenfe than thofe of the common fort of Peruvian bark.

The infufion, we are told, bears to be mixed in any proportion with fpirits, without becoming turbid, or producing any kind of decompofition. The firft decoction is confiderably deeper-coloured than the infufion (which colour it retains in paffing the filter), and poffeffes the fame powers in a higher degree, but does not retain them fo long, nor is it fo pleafant to the tafte. On ftanding any length of time with the chalybeate, the colour becomes pale, and is fooner decompofed than the cold infufion: on ftanding fome days it lets fall a fmall quantity of a reddifh, earthy fecula, which is intenfely bitter and aftringent; the fuperincumbent liquor becoming gradually clearer and clearer, and at the fame time of a deeper red, much refembling the tincture. Lime-water added to the decoction,

infufion,

infufion, or diluted tincture, darkened them
confiderably, and caufed in all a copious depo-
fition of reddifh brown fecula, which became
purple coloured in twenty four hours. The decoction, it is obferved, gave the greateft quantity
of fecula. An infufion of pale Peruvian bark,
prepared in every refpect the fame as the infufion
of Swietenia bark, was treated with lime-water in
the fame manner, and formed a feparation, but
in a much lefs degree.

Bark of *Melia Azadirachta* (Margofa tree) treated exactly in the fame manner, formed a feparation of a lighter-coloured fecula, in a much
greater quantity than the common Peruvian
bark, but much lefs than the Swietenia bark.

The clear reddifh-coloured liquor, we are told,
that floats over the precipitate caufed by the addition of lime water, is void of aftringency to the
tafte, or has it only in a trifling degree ; but for a
farther proof, it feems, a chalybeate was employed, which did not in the leaft darken this liquor;
but a greenifh tinge was produced, together with
a further decompofition and precipitation of a
reddifh fecula. This experiment, our author
thinks, ferves to fhow that at leaft the aftringent
part of the bark is entirely thrown down by
lime-water; and he confidered this as fo intereft-

ing

ing a point, that he repeated the fame experiment with this, as well as with other aftringent barks, and found the refult exactly the fame.

The fame chalybeate added to lime-water of the fame ftrength as that employed in the abovementioned experiments, produced a fmall, green cloud ; the Swietenia bark infufion thrown into this produced a muddinefs, and foon after, a copious precipitation of dirty-coloured fecula.

An infufion of this bark in lime water is deeper coloured than the plain infufion, but poffeffes very little bitternefs, and ftill lefs aftringency. A chalybeate added to this infufion rendered its red colour a little deeper only, and no decompofition took place : after ftanding fome time, the infufion had no tafte of the lime-water.

From thefe experiments, Dr. Roxburgh confiders lime-water as a very improper addition ; and obferves that, in this refpect, they agree with thofe made by Dr. Irving on the red and quilled Peruvian barks.

Vitriolic acid rendered the firft decoction, or watery infufion, paler ; and, upon ftanding, it became a little turbid, and let fall a fmall quantity of a light-brown fediment.

Vinegar had the fame effect.

Mild, or cauftic vegetable alkali, or mild

foffil

foffil alkali, foon deepened and rendered brighter the cold watery infufion or decoction, nor did any decompofition take place in forty-eight hours.

Mild magnefia, fimply added, rendered the colour of the infufion paler, without fenfibly altering the tafte.

Alum has been at times fuccefsfully employed for the cure of intermitting fevers, and the analogy it bears to other tonics renders it a likely remedy. Our author was therefore defirous to try what would take place on adding it in a fmall quantity to infufions and decoctions of this bark. The addition, it feems, rendered their colour paler, and a little decompofition took place, with a precipitation of a fmall quantity of a light-brown fecula: to the tafte it increafed the aftringency without fenfibly diminifhing the bitter; but with alum they did not change their colour when a folution of green vitriol was added.

Eight ounces of the coarfe powder were boiled in ten pints of foft well water to four pints; the refiduum was repeatedly boiled in frefh parcels of water, exactly in the fame manner for eleven times, when the liquor

K 4   came

came off ftill much coloured, but taftelefs, and
fhowed no figns of aftringency with the chalybeate; the tenth decoction excepted, which did
fhow figns of aftringency, as it was darkened a
little by it.

The frefh decoction of common Peruvian
bark, made fimilarly, but in a fmaller quantity, ftruck flowly about as deep a colour with
the fame chalybeate, as the fourth or fifth decoction of Swietenia bark did quickly.

As the eleventh decoction was taftelefs, although coloured, it was thrown away; the other
ten had been regularly ftrained, while hot, and
fuffered to ftand till perfectly cold, then poured
off, clear from fediment; they were mixed, and
evaporated to a hard extract, which weighed
two ounces and three-quarters. The extract,
when foft, was of a dark red colour, flavourlefs,
fmooth, homogeneous, and unctuous when rub-
bed between the fingers and thumb. The tafte
of the decoction was well preferved in this extract; the moft minute part of it, diffolved in water, ftruck a black colour with martial folution as
quickly and as deep as the decoction itfelf, but
the tafte was not fo ftrong as might be expected
from that of the bark. This, our author thinks,
might

might perhaps be owing to the more fixed, inert parts, extracted by the long and repeated boilings (which lasted two days) being mixed in the mass of extract. But this, he observes, would not be the case, or but in a small degree, with one prepared from only one or two boilings. To determine this point, he boiled one ounce of the powdered bark in two pints of water, pretty briskly, down to one pint; after the liquor was poured off, to the residuum were added two other pints of water, and boiled in the same manner. The decoctions were mixed, and evaporated to a dry extract, which weighed two drachms and a half, and was in taste, &c much as the former from ten coctions; the proportion of extract from two boilings is therefore, he observes, nearly equal to that of ten: so that, although the decoctions were highly coloured, and considerably bitter and astringent, even to the tenth, yet they could have contained but a small portion of the powerful qualities of the bark.

The residuum, when perfectly dry, weighed four drachms and a half; and spirit of wine being poured on it, though assisted at times with the heat of the sun for many days, extracted
neither

3

neither colour nor taſte, ſo completely had the
virtues of the bark been extracted by the water.

Dr. Roxburgh obſerves that the dry extract
imbibes much moiſture when the weather is
damp; ſo much as to make it ſtain the fingers, or
any thing that touches it : that it melts readily
in the mouth ; is eaſily ſoluble in water and in
ſpirits ; and, like the decoction and tincture,
bears to be mixed without decompoſition.
Theſe ſolutions and mixtures, we are told, re-
ſemble much the original decoction and tincture, and their mixtures, both in taſte and colour.

Should this ever become the valuable drug it
promiſes, it would be adviſable, our author
thinks, to have the extract prepared on or near
the ſpot where the trees grow. If this is done
during the hot ſeaſon, the evaporation, he obſerves, might be effected by the heat of the ſun
and hot winds, which would certainly produce
a much more elegant, efficacious extract than
could poſſibly be prepared in any other way or
place, and would alſo preclude every idea or
chance of its being ſophiſticated.

This bark, he finds, contains much mucilaginous matter, the cloth that the decoctions

were -

were ſtrained through, having become, when dry, ſtiff as if ſtarched. This, he thinks, may account for the decoctions remaining ſo many days turbid, which is, no doubt, he adds, favourable for the action of the ſtomach upon the bark. The late Dr. Fothergill, he obſeives, recommended an addition of ſome mucilage to decoctions of common bark, in order to keep them turbid, that the active parts might be kept more completely ſuſpended in the liquor *.

In the way of diſtillation, this bark, it ſeems, yields nothing, not the ſmalleſt apparent quality, either with water or ſpirits. In this reſpect, Dr. Roxburgh thinks, it reſembles exactly both the pale and red Peruvian barks, viz. in having its powers or virtues of a very fixed nature.

Rectified ſpirit of wine extracts from the bark a clear, deep red tincture, poſſeſſing the aſtringency of the watery infuſion or decoction, and more of the bitter. If not too ſtrong, it makes, we are told, one of the moſt pleaſant bitters we are in poſſeſſion of; and it bears to be diluted with water in any proportion, without decom-

---

* Med. Obſ. and Inq. Vol. I. p. 321. 2d Edit. 8vo. London, 1758.

poſition,

pofition, which renders it in many cafes the more defirable.

Four ounces of powdered bark were infufed, by our author, for eight days, in three pints of French brandy; thefe were poured off, and four pints more of the fame brandy added, which, after ftanding four days, were alfo poured off: both thefe infufions were mixed, and he drew off, by diftillation, a quantity of the fpirit, which (as before obferved) did not in the leaft partake of any of the qualities of the bark : the reft was gently evaporated to a dry extract, which weighed nine drachms. The extract itfelf was of a much darker colour than that procured by water, and was dried with more difficulty ; but the tafte of the two extracts was much the fame. The refiduum was boiled in fix pints of water to two, and the decoction was found to be ftill pretty ftrong to the tafte, both in bitternefs and aftringency. This induced him to repeat the boiling, twice more, with frefh parcels of water ; and the laft decoction, though weak, was ftill bitter, and fhowed figns of aftringency, with a martial folution. Thefe four decoctions were mixed and evaporated to a dry extract, weighing three drachms, which added to the fpirituous ex-

tract,

tract, made in all twelve drachms, from four ounces of powdered bark, and agreed nearly with the quantity procured by water alone.

The antiseptic powers of this bark, according to our author's experiments, are not inferior to its bitter and astringent qualities; for watery infusions in open phials kept perfectly good for sixty days, without any tendency to fermentation, except a few air bubbles, which they discharged about the second day; indeed they acquired strength, we are told, as the colour produced at the end of that time (sixty days), by the addition of a chalybeate, was darker, and as instantaneous as at any prior period.

Sixty grains of the lean of raw mutton were preserved sweeter and firmer in an infusion of ten grains of this bark in four ounces of water, than an equal quantity of the same mutton in a similar infusion of pale quilled Peruvian bark. The flesh was tinged red by the infusion of Swietenia bark, and its fibres were firm and distinct at the end of twelve days; while that preserved in the Peruvian infusion was white, and its fibres softer, and infinitely more fetid.

Almost all the foregoing experiments, it is observed,

obferved, were made firft with bark of the fmaller branches, and again with bark of the trunk of a large tree; the latter was evidently ftrongeft.

The feeds of this tree are defcribed as a ftrong, fimple, pleafant b⬤r, without any of the aftringent power. The leaves poffefs nearly if not all the aftringency of the bark, and a very large proportion of its bitter ; but their tafte is faid to be not fo agreeable either in fubftance or in infufion.

From the foregoing analyfis, Dr. Roxburgh ventures to draw the following conclufions :

Firft. That the active parts of the bark of this fpecies of Swietenia are much more foluble than thofe of Peruvian bark, particularly in watery menftruums.

Secondly. That it contains a much larger proportion of active (bitter and aftringent) powers, than Peruvian bark.

Thirdly. That the watery preparations of this bark remain good much longer than fimilar preparations of Peruvian bark.

Fourthly. That the fpirituous and watery preparations bear being mixed in any proportion, without decompofition.

Fifthly. That the bark in powder, and its
preparations,

preparations, are much more antifeptic than
Peruvian bark, or fimilar preparations of it.

Now, fince this bark yields fo much of its
virtues to cold water, as to preferve flefh from
corruption, in a hot climate, with the thermometer from 87° to 102°, it is reafonable, he
contends, to fuppofe it will yield ftill more of
its tonic and antifeptic virtues in the ftomach,
where it meets with the moft powerful folvents :
we have therefore, he thinks, much to expect
from it in the cure of gangrene and other putrid difeafes.

Bitters and aftringents, in a feparate ftate, our
author obferves, are confidered as tonic remedies; but when found combined in the fame
fubftance, they become ftill more powerful : it
is from thefe qualities, he contends, that the
beft judges allow the Peruvian bark to derive
its virtues. On this point he quotes the authority of Dr. Cullen, who has remarked, " that the
" recurrence of the paroxyfms, in intermitting
" and remitting fevers, depends on the recur-
" rence of atony in the extremities of the arterial
" fyftem; hence they are prevented by fuch
" tonic medicines as obviate this atony : a
" great variety of aftringents and fimple bitters
" have been found to anfwer that end, but
" none, hitherto difcovered, fo effectually as the
" Peruvian

" Peruvian bark, on account, it is thought, of
" its poffeffing thofe powers conjoined *."

The antifeptic qualities of Peruvian bark, our
author obferves, are alfo great; hence its ufe in
the cure of all febrile putrefcent diforders, attended with debility, putrid ulcers, &c.

From the evident qualities of this new bark,
and from the fuccefsful experience he has had
with it, in intermittent fevers†, &c. Dr. Roxburgh

---

* Treatife on the Materia Medica.

* Hiftories of feveral of thefe cafes have been communicated by Dr. Roxburgh to the College of Phyficians at Edinburgh, and an account of them is given by Dr. Duncan in the
differtation referred to in a former note, together with the refults of feveral trials made with this bark, by his father, in
the Clinical Ward of the Royal Infirmary at Edinburgh. We
fhall take the liberty of tranfcribing this part of his work:

" Morbus, quo Roxburgius hunc corticem fæpiffimè adhi-
" bendam curavit, febris quotidiana apud Cullenum nun-
" cupatur. Rariùs ex toto, fed ex parte, et ad breve tantum-
" modo tempus, remittens, periculofiffimus erat. Ægroti ferè
" omnes hoc morbo correpti fuerant, dum incolebant iftos
" montes ingentes, qui Indiæ peninfulam tranfcurrunt. Inter
" hos montes fylvæ opacæ, denfa ferarum tecta, convalles pa-
" ludofas, hominum generi peftiferas, ubique obumbrant. Se-
" des eft indigenis etiam, confuetudine licet obfirmatis, infa-
" lubris, advenis autem adeo perniciofa ut pauci, perpauci
" quidem, quos dira neceffitas inter hos montes hiemare coege-
" rit, morbo hoc atrociffimo immunes fint. Tali febre, tali
" tempeftate

burgh has every reafon to imagine it will prove
equal, if not fuperior, to the Peruvian bark, for
every purpofe for which that medicine is ufed.

Our

" tempeftate laborantium ne dimidiam quidem partem con-
" valefcere Roxburgius affimat.

" Cal. Junii, A. D. 1791. Indus annos natus viginti,
" habitûs tenuis, nonnullis antè menfibus, dum prope montes
" occupabatur, febre quotidianâ affeƈtus erat. Corticem
" Cinchonæ officinalis aliquantifper fine fruƈtu affumpferat;
" idcirco Roxburgius, et quia ipfe parvas corticis Soymidæ
" quantitates impunè adhibuerat, ægro nihil à periculo ab-
" horrenti grana viginti pulveris ex aquæ cyatho fumenda
" præfcripfit. Duabus exinde horis, fcrupuli duo adhibiti
" funt; et, poft fimile temporis intervallum, drachma. Cor-
" tex ægro nequaquam ingratus erat, alvumque folvit. Æger,
" cortice poftea ad drachmam, unaquâque intermiffione, af-
" fumpto, triduo febre immunis erat.

" Pridie Iduûm Augufti, A. D. 1791, J— V— Lufi-
" tanus *, annum agens quadragefinum quintum, ejufque duæ
" filiæ, altera fex, altera tres annos nata, manferant aliquan-
" diu, inter menfem proximè præteritum, intra montium ter-
" minos; initioque menfis labentis, febre quotidianâ, quæ
" nihil fermè quicquam remifit, affeƈti funt. Febre remit-
" tente, femper alterâ quaque horâ fumebant aquæ ex cortice
" Soymidæ* pater fefcunciam, filia major natu unciam, et
" minor femunciam. Duos poft dies, à morbo valebant.

＊ " R. pulv. cort. Swiet. Soymidæ unciam unam,
          " aquæ fontanæ libras duas.
    "Mifceantur, et phialâ priús agitatâ, modo præfcripto fumantur."

* Vide p. 148.

Our author next enumerates different fpecies of Cinchona, viz.

Firft.

" Morbus, quo hi quatuor ægroti laborabant, partim ob
" anni tempus, quo febre correpti funt, atque partim ob
" tempeftatis ficcitatem, folito levior erat; atque Roxbur-
" gius, propter ægrorum debilitatem, neque evacuantia ad-
" hibebat, nec intermiffiones expectabat.

" xv. Cal. Sept. A. D. 1791, B— Lufitana, habitûs
" infirmi, nonnullos dies, febre gravi, nunquam ex toto re-
" mittente, laboraverat. 'Antimonium tartarifatum ex multâ
" aquâ, partitis vicibus, ufque ad vomitionem, adhibuit.
" Poftero die drachma corticis Soymidæ, in remiffione mi-
" nimè adhuc notabili, ter affumpta eft. Intermiffio proxima
" plenior evafit, atque, ex corticis ufu, biduo poftea morbus
" ipfe, fimulque diarrhœa quâ laboraverat ægra, ceffârunt.

" Menfe Septembris, A. D. 1791, J. E— decurio Euro-
" pæus, annos natus quadraginta, febre remittente graviter
" affectus eft. Receffus principio ferè nulli, ex ufu præpara-
" torum ex antimonio notabiliores evaferunt; et æger, quan-
" quam omni generi intemperantiæ deditus, cortice ter fin-
" gulis intermiffionibus adhibito, paucis diebus convaluit.

" vii. Cal. Sept. A. D. 1791, T. L— annos natus
" octodecim, quofdam dies febre biliofâ laboraverat; cujus
" receffus, etiam poft antimonii tartarifati ufum, parum
" notabiles erant. Debilitate autem urgente, fcrupuli duo
" corticis Soymidæ, omni receffu, ter adhibebantur, et, ad
" alvum folvendam, lixiva tartarifata.

" A cortice autem nihil proficiente, iii. Cal. deceffum
" eft; atque medicamentis idoneis affumptis, febris prorfûs
" fere,

Firſt. *Cinchona Officinalis panicula brachiata;*
to this ſpecies, he obſerves, belong the pale,
quilled,

" ſere, ſtatis temporibus, intermiſit. Soymida nunc iterum
" adhibita, quatuordecim diebus, morbum penitùs fugavit.

" Menſe Septembris, A. D. 1791. S— nutrix laɛtans,
" annos nata triginta quinque, febre quotidianâ correpta
" eſt. Alvo, inter primam intermiſſionem, ſodâ vitriolatâ
" ſolutâ, morbus triduo cortice Soymidæ depulſus eſt; ſed
" lac interim fluere ceſſaverat.

" Menſe Septembris, A. D. 1791. Indus, ſervus domeſ-
" ticus †, febre ſingulariter intermittente ægrotavit. Sub
" occaſum ſolis, acceſſit febris gravis, quæ horâ nonâ veſ-
" pertinâ intermiſit. Oriente autem ſole, iterum acceſſit,
" atque, horam circiter nonam ante meridiem, denuo inter-
" mittens, ægrum viribus integrum reliquit. Exinde cor-
" tice Soymidæ ter, ſingulis intermiſſionibus matutinis, ad
" ſcrupulos duos adhibito, triduo morbus omnino evanuit.

" J— R— Europæus annum agens trigeſimum, vitio pul-
" monis multum debilitatus, ineunte Oɛtobri febre quotidi-
" anâ, cui erant acceſſiones veſpertinæ, affeɛtus eſt. Tertiâ
" intermiſſione, duo corticis ſcrupuli bis adhibiti alvum
" magnopere ſolverunt. Soymidâ nihilominus continuatâ,
" æger quatuor diebus à febre valebat.

" Pridie Iduûm Decemb. R— miles Indicus, annos natus
" triginta, febre quotidianâ tredecim dies, medicamentis
" vernaculis nihil proficientibus, laboraverat. Intermiſſione
" proximâ duo corticis Soymidæ ſcrupuli ex aquâ ter adhi-
" biti alvum bis cierunt, morbumque levarunt. Cortex
" repetitus ægro ſanitatem reſtituit.

" Pridie Iduûm, Dec. A. D. 1791, L— miles Indicus.

† Vide p. 148.

" annos

quilled, and red barks, which the beſt judges
imagine are from the ſame tree; the thick-
red

" annos natus viginti tres, antecedente die, febre quotidianâ
" affectus eſt. Cortice ter ſingulis intermiſſionibus adhibito,
" alvus ſoluta eſt, morbuſque mox remiſit.

" Pridie Iduûm Dec. S. N— miles Indicus, annos natus
" quadraginta, ıv. Non. Decemb. febre correptus erat. Nullis
" hactenus medicamentis uſus, magis nunc magiſque debilis
" evaſerat. Cortex in remiſſione ter adhibitus ventrem
" ſolvit, triduoque morbum depulſit.

" Pridie Iduûm Dec. A. D. 1791. N— miles Indicus,
" annum agens vigeſimum quintum, pridie febre quotidianâ
" affectus erat. Cortex Soymidæ, ter in unaquaque inter-
" miſſione adhibitus, alvum movit, atque morbum brevì ſu-
" peravit.

" vııı. Cal. Martii, A. D. 1792. J. V— per biduum
" febre iterum *. laboraverat. Morbo autem duo acceſſus
" totidemque remiſſiones quotidie erant, ejus inſtar paulò
" ſuprà deſcriptæ †. Cortex Soymidæ, in matutinis inter-
" miſſionibus, alterâ quaque horà adhibitus, triduo febrem
" curavit.

" Circiter medium Februarii, R— infector telæ xylinæ,
" annos natus viginti quinque, laborans tumore hypogaſtrii
" æquali, dolente, quem comitata eſt febris omni mane rece-
" dens, atque alvus aſtricta, ad Roxburgium adductus eſt;
" cui dixit, ſe duodecim antè dies affectum eſſe dolore circa
" umbilicum torquente, qui uno alterove die gravis evaſit,
" atque profundus, et, quaſi inter veſicæ urinariæ fundum
" atque inteſtinum rectum, ſedem cepit; abdomen mox tu-
" muiſſe, ipſumque toto corpore febricitâſſe; cauſam autem

* Vide p. 145. † Vide p. 147.

" ignorâſſe

red fort being from the trunk, while the pale-
quilled fort is from the branches, and from
young

" ignoraffe malorum; multa denique remedia vernacula
" incalfùm adhibuiffe.

" Ei præcepit medicus, ut affumeret parvas lixiviæ tar-
" tarifatæ quantitates, donec fuperveniret catharfis, pro
" potu communi biberet aquam ex tamarindis coctam cum
" faccharo, et ut interea diætà ex oryzâ famem tolleret.

" Alvo his exoneratâ, meliufcule fe habere fenfit æger;
" tumori autem nequaquam decrefcenti, veficatorium ad-
" motum eft, alvufque lixivâ tartarifatâ et aquâ ex tama-
" rindis cum faccharo commiftâ foluta eft.

" Per noctem febris invaluit. Die autem, à curatione
" inceptâ, tertio alvus vehementer fluxit. Dejectiones
" purulentæ admodum erant, peffimè olentes, colore per-
" virides. Tumor ftatim subfedit.

" Æger maximè debilitatus, per noctem, graviter fe-
" bricitabat. Mane igitur, cùm primùm febris fe remifif-
" fet, ei pulvis ex Soymidæ cortice et lixivâ tartarifatâ com-
" pofitus adhibitus eft, et, die progrediente, ter repetitus.
" His factis, alvus purulenta quædam quater dejecit. Hâc
" curatione triduo pòft à febre valebat, et, cortice nunc
" femel tantum in die adhibito, decem diebus domum re-
" diit fanus.

" Roxburgius unam tantum occafionem corticis Soymidæ
" contra gangrænam adhibendi nactus eft. Viro diffoluto,
" per idem tempus lue Venereâ laboranti, fuper mediam
" tibiam ulcus erat. Cùm Soymidæ pulvis eius ftomacho
" nigratus effet, extracto ufus eft, atque, expectatione ci-
" tiùs, morbo immunis evafit. Perhibet præterea Roxbur-

" gius,

young trees. The Spaniards themfelves, he
adds, employ the red fort.

Second.

" gius, Duffinum chirurgum valetudinarii Madrafienfis pri-
" marium hunc corticem contra iftiufmodi mala maximo
" cum fructu adhibuiffe.

" His memoratis, Roxburgius ingenuè fatetur infignem
" tempeftatis ficcitatem, hujus novæ Swieteniæ corticis
" ufum feliciorem forfitan reddidiffe. Notat præterea, cor-
" ticem primo die alvum plerumque moviffe, poftea autem
" nunquam, neque profectò, præter morbi curationem,
" ullos ex ejus ufu effectus obfervâffe. Cur, ante corticis
" ufum, non fæpiùs, ut mos plerifque eft, vomitum et al-
" vum moviffet, hanc rationem reddit, nempè ex regionis
" naturâ, ex victu, ex vitâ, atque ex religione, corpora
" Indis effe gracilia, nec plena, ne inflammationibus ob-
" noxia; atque remediis, quæ ante corticem adhiberi fo-
" lent, febres, ut ille putat, in longum fæpe trahi, et iis,
" æque ac morbo fere ipfo, ægrotos infirmari.

" Hæc uberiùs dixi atque fufiùs eò quòd ex his potiffi-
" mum, quantum polleat hic cortex, apparet. His adduc-
" tus pater meus, cùm ægrotos nofocomio Edinburgenti
" curabat, atque difcentibus de iis prælegebat, nova hujus
" corticis tentamina facere voluit. Hâc autem regione,
" cùm febris intermittens perrara fit, nobis nulla, quid pro-
" ficiat cortex nofter, experiendi idonea fatis occafio oblata
" eft. Nonnullis autem ægrotis adhibita eft.

" XIII. Cal. Jan. A. D. 1793. Joannes M'Kay, annum
" agens vigefimum tertium, priufquam in nofocomio recep-
" tus erat, duodecim dies febre, cujus acceffiones altero
" quoque die redibant, laboraverat. Sed, cùm, ab initio

" horror

Second. *Cinchona Caribæa*; the Caribæan
or

" horror et calor per idem tempus duraviſſent, ſudor pror-
" sùs defeciſlet, atque mala pectoris, coma, et torpor fe-
" brem comitata eſſent, hæc affectio minimè idonea, in quam
" novum medicamen tentaretur, videbatur. Cortéx igitur
" Cinchonæ rubræ, per duodecim dies adhibitus eſt; cùm
" autem acceſſus poſt intervalla, licet valde diſlimilia, ad-
" huc redirent, ægro, ut Soymidæ drachmam alterâ qua-
" que horâ ſumeret, præſcriptum eſt. Alvum torminibus
" magnopere movit, acceſſus autem proximus poſtremus
" erat. Convaluit.

" Jacobus Grant, annos natus viginti quinque, qui ali-
" quandiu in noſocomio propter teſtis tumorem manſerat,
" viii Iduûm Junii, A. D. 1793, herbâ humidâ veſperi
" recumbens, frigore, gravi dyſpnœa atque anguſtiæ in
" faucibus ſenſu, affectus eſt. Hæc facile ætheri vitriolico
" ceſſerunt, cortexque Cinchonæ, quo vires proſtratas re-
" ficeret, adhibitus eſt. v. Iduûm iterum frigore, dyſpnœâ,
" atque vomitione ſanguinolentâ, correptus eſt. Quinto
" poſtea veſpere horrores, intermittentis inſtar, acceſſerunt.
" Uſum corticis Cinchonæ, quippe qui acceſſionibus nihil
" obſtaret, intermiſit medicus, pulveremque corticis Soy-
" midæ, duplici autem quantitate, in ejus locum adhibuit.
" Hoc facto morbus nunquam poſtea rediit.

" Duabus adoleſcentulis, alteri à ſingulari affectione hyſ-
" tericâ, convaleſcentibus cortex Swieteniæ Soymidæ, ut
" corpora firmaret, ſi non cum utilitate ſaltem ſine incom-
" modo, adhibitus eſt.

" Vi inſuper aſtrictoriâ pollere, ſatis conſtat è muliere
" annorum quadraginta ſex, quæ leucorrhœâ laborabat.

L 4   Duobus

or Jamaica bark of Dr. Wright *. This laſt, our author obſerves, poſſeſſes in a higher degree the bitter, but is very weak in the aſtringent power, and ought not to be depended on when the other is procurable.

Third. *Cinchona Sanctæ Luciæ, floribus paniculatis, glabris, laciniis linearibus tubo longioribus, ſtaminibus exertis, foliis ellipticis glabris*; Saint Lucia, or new bark. This is another ſort, which has been introduced into practice : but its being poſſeſſed of ſtrong emetic and purgative qualities, renders it, in our author's opinion, leſs eligible, particularly after the paſſages have been cleared. Theſe properties, he obſerves, the Jamaica bark does not poſſeſs ; which eſtabliſhes a ſtriking difference.

Fourth. *Cinchona Corymbifera, foliis oblon-*

" Duobus ſenibus ventris fluxu affectis nihil profecit. Hi " autem, omnia, quæ alvum aſtringunt, experti, morbo " non levato, è noſocomio egreſſi ſunt.

" Cortex Soymidæ, ut multum, necne, contra putredi- " nem poſſet, appareret, quinque ægrotis typho putrido " laborantibus adhibitus eſt. Omnes convaluere. His " ventrem adeo non movit, ut, per totum morbum, alvum " aliis medicamentis ducere opus eſſet." *Vide* Duncan *Tentam. de Swietenia Soymida*, p. 41. et ſeq.—EDITOR.

* See Philoſ. Tranſact. Vol. LXVII. page 504; and London Medical Journal, Vol. VIII. page 239.

gis,

gis, lanceolatis, corymbis axillaribus; of Dr. Forſter; is a native of the South-Sea Iſlands: but of its virtues we know nothing more, than that he ſays, " it is like Peruvian bark, bitter " and aſtringent."

Fifth. *Cinchona Orixenſis, foliis oppoſitis, tomentoſis, ſtipulis interfoliaccis, ſemilanceolatis, floribus terminalibus, paniculatis, tomentoſis, capſula valvis contrariis à vertice dehiſcens*; of Dr. Roxburgh. The ſtructure of the capſule, he obſerves, forms the chief difference between this and Cinchona Officinalis, for the ſeeds are exactly as delineated by Gærtner, and the reſt of the definition correſponds with that given by Linnæus. It is a native of that chain of mountains which ſeparates the northern provinces, or circars, from the Mahrattah dominions immediately behind them. The bark of this ſpecies likewiſe is bitter and aſtringent.

Dr. Roxburgh has alſo found another new ſpecies of Swietenia, a middle-ſized tree, the wood of which is very heavy, cloſe-grained, and yellow; the bark likewiſe is yellow, and very bitter, but poſſeſſes much leſs aſtringency than that of the S. febrifuga, and its aſtringengy, he obſerves, is of a peculiar kind, for the colour produced, on an infuſion, with a martial ſolution, was a dark brown.

There

There is alfo the bark of another large
tree, which, at the time of writing this account,
he tells us, he had under examination, and
which is likewife very bitter : the Hindoos
call it *Wallurfe.* It will, he imagines, form a
new genus in the clafs Decandria, and order
Monogynia. Its effential characters are *calyx
quinquefidus, petala quinque, nectarium duplex,
exterius cylindricum oré decemfido, antheras gerens,
interius annularium, bafin germinis cingens, bacca
monofperma.*

The bark of this tree, we are told, is in high
repute as a medicine amongft the Hindoo phyficians; and gives name to a compound foft extract, called *Walluvodufay,* which they employ in a variety of difeafes.

It alfo poffeffes powers of a very different
nature; for, powdered and thrown into pools
where there are fifh, it foon intoxicates them
to that degree, that they are eafily taken with
the hand.

Dr. Roxburgh obferves that the bark of
Melïa Azadirachta, already taken notice of*,
has frequently and fuccefsfully been employed
as a fubftitute for Peruvian bark, in the cure

---

* Page 133.

3

of

of remittents and intermittents; and that an infufion or decoction of its leaves is alfo a good anthelmintic, and as fuch employed by the Hindoos.

The bark of another large tree, which our author calls *Nauclea Daduga*, poffeffes alfo, he tells us, in a confiderable degree, both the bitternefs and aftringency of Peruvian bark; and he thinks it is next in power to that of the Swietenia febrifuga. Although this tree differs widely in its flower from the hitherto known fpecies of Cinchona, yet in its parts of fructification it agrees with them, it feems, almoft exactly.

X. *An*

X. *An Account of the Effects of Mahogany Wood in Cases of Diarrhœa. By* Mr. Francis Hughes, *Surgeon of the General Infirmary at* Stafford. *Communicated in a Letter to* Mr. John Pearson, *Surgeon of the Lock Hospital, and by him to Dr.* Simmons.

AN accidental circumstance first suggested to me the idea that mahogany wood might prove serviceable as a medicine; for I did not then know that any part of the tree had been employed for medicinal purposes. I was accordingly induced to make use of it in cases of diarrhœa, both in decoction and in the form of an extract; and after repeated trials, I can venture to assert that I have not been disappointed in the expectations I had formed of its efficacy.

For the decoction I boil an ounce of the shavings of Jamaica mahogany wood in two pints of water, till one pint of the liquor is wasted, and then strain off the remainder for use.

The extract I make use of has been prepared
by

by boiling the fhavings of the fame wood in repeated affufions of frefh water, in the fame proportion and manner as are directed for the extract of logwood *(extractum hæmatoxyli)* of the London Pharmacopœia. The quantity of extract obtained in this way amounts to fomething more than ⅛ of the fhavings employed. The Honduras mahogany wood is of a paler colour, and lefs aftringent than the Jamaica, and does not yield quite $\frac{1}{10}$ part of extract.

Both the decoction and extract are very bitter and aftringent, leaving a roughnefs in the mouth for fome time after they have been tafted.

The extract, in its appearance, refembles gum kino. It diffolves completely in water, and in fpirit of wine, and ftrikes a black colour with falt of fteel.

The following are fome of the Cafes in which I have employed thefe remedies.

CASE I.

In July, 1793, a foldier belonging to a regiment on the Irifh eftablifhment, who is a native of Stafford, was fent hither from his regiment

giment for the recovery of his health. He had for fome time been unfit for duty, and was much reduced by a diarrhœa, which having come on after a fever, had continued feveral months, and refifted a variety of medicines.

I gave him an ounce of the decoction three times a day, and as it fat eafy on his ftomach, and feemed to have a good effect, the dofe, after the third day, was increafed to an ounce and a half. He perfevered in the ufe of it during fixteen days; the diarrhœa gradually fubfided; his appetite and ftrength returned; and at the end of that time he was fufficiently recovered to go back to his regiment in Ireland.

CASE II.

A woman of a thin, delicate habit, applied to me in October, 1793, on account of a violent diarrhœa, for which fhe had taken different medicines without any good effect. It had come on, fhe faid, after fitting up a whole night in wet clothes, and had continued more than a fortnight; fhe was free from fever.

I directed her to take pills compofed of fix

grains

grains of the extract, three times a day. Within the space of a week the diarrhœa was much abated, and she had acquired strength ; she persevered, however, in the use of the medicine for the space of three weeks, at the end of which time the complaint had entirely ceafed. A fluor albus, with which she had been troubled many months, was likewise much abated ; but perhaps this latter circumstance ought rather to be ascribed to the improved state of her general health, than to any specific effect of the medicine.

CASE III.

In January, 1794, I was applied to by a man fifty years old, who for several years had been a hard drinker, and was now extremely emaciated ; his legs were oedematous ; he had no appetite ; was subject to frequent vomiting, and had a slight diarrhœa.

I gave him aromatic bitters for several days, but finding no amendment, I determined to have recourse to the mahogany. I gave him eight grains of the extract, made into pills, three times a day. At the end of five days his

disposition

difpofition to vomit had ceafed, and he had a little appetite. He continued the ufe of the medicine for ten days longer, and was then fo much relieved as to be able to walk and ride out every day. This ftate of amendment continued for a fortnight, when he relapfed into his old habit of drinking, and his former fymptoms returned. Recourfe was again had to the fame medicine, but without effect.

To the above I could add many other in-ftances of the good effects of the extract and decoction in cafes of long continued diarrhœa, where the complaint feemed to depend on a morbid irritability of the ftomach and inteftines, and where the ufe of tonic and aftringent me-dicines appeared to be indicated. The few hif-tories I have related will, I truft, be fufficient to point out the modes of adminiftering the re-medies in queftion, and the effects that may be expected from them; and perhaps will induce medical practitioners to extend a trial of their efficacy to other difeafes.

The dofes in which I have hitherto given thefe remedies have been fmall; but much larger dofes may be given with fafety, and in many cafes will, I am perfuaded, be more efficacious.

To

To try the effect of a confiderable dofe on the ftomach, I took two ounces of a decoction, prepared by boiling two ounces of the fhavings in two pints of water to a pint, which is twice the ftrength of the decoction defcribed in Cafe I. and which I have ufually adminiftered. At firft I perceived no effect from it; but at the end of ten minutes a difagreeable naufea came on, with a flight pain at the ftomach, and a glowing fenfation fimilar to that produced by the taking a glafs of ftrong wine. Thefe effects gradually went off in about half an hour, and I felt no other inconvenience from the dofe.

Stafford, Feb. 12, 1794.

XI. *Account of some Discoveries made by Mr. Galvani, of Bologna; with Experiments and Observations on them.   In two Letters* from Mr.* Alexander Volta, F. R. S. Professor of Natural Philosophy in the University of Pavia, to Mr. Tiberius Cavallo, F. R. S.—From the Philosophical Transactions of the Royal Society of London, for the Year* 1793. *Part* I. 4to. London, 1793.

THE subject of the discoveries and researches, concerning which I am about to write to you, Sir, is *animal electricity*; a subject which cannot but be extremely interesting to you.   I know not if you have yet seen the work of a Professor of Bologna, Mr. Galvani, which appeared about a year since, with this title; ALOYSII GALVANI *de Viribus Electricitatis in Motu Musculari Commentarius.* Bononiæ, 1791, in 58 pages, 4to, with four large plates; or at least if you have had any

* In the Philosophical Transactions these two letters are given in French; for the present translation of them we are indebted to the kindness of a friend.—EDITOR.

account

account of it*. It contains one of the moſt
beautiful and ſurpriſing diſcoveries, and the
germe of ſeveral others. Extracts from this
work have appeared in different Italian Jour-
nals, and, among others, in that entitled *Gio-
nale Fiſico-medico*, publiſhed by Dr. Brugna-
telli, of Pavia, to whom I myſelf have ſent two
long papers, which will be followed by ſeveral
others, as I have conſiderably extended my experiments and inquiries on this ſubject. The
letters I now addreſs to you are intended as
a ſketch both of the admirable diſcovery of
Mr. Galvani, and of the progreſs which I have
been fortunate enough to make in this new path;
and I requeſt you, Sir, to preſent them to Sir
Joſeph Banks, Bart. the worthy Preſident of the
Royal Society, to be communicated, if he
thinks proper, to that learned body, as a feeble
teſtimony of my gratitude for the honour they
have done me in electing me one of their number, and of my zeal and eagerneſs to comply
with their invitation to communicate to them,
from time to time, the fruit of my reſearches.

(1.) Mr. Galvani having diſſected and prepared a frog, in ſuch a manner that the legs remained attached to a part of the back bone,

* See Vol. III. p. 180.—EDITOR.

ſeparated

separated from the reft of the body, folely by
the crural nerves, which were laid bare, obferved that very lively motions were excited in
thefe legs, with fpafmodic contractions in all
the mufcles, every time that (this part of the
animal being placed at a confiderable diftance
from the conductor of an electrical machine,
and under certain circumftances, which I fhall
explain hereafter) a fpark was drawn from this
conductor, not on the body of the animal, but
on any other body, or in any other direction.
The requifite circumftances, therefore, were,
that the animal thus diffected fhould be in contact with, or very near fome metal or other
good conductor, of fufficient extent, or, what
was ftill better, between two fimilar conductors, one of which fhould be turned towards
the extremities of the legs of the animal, or
fome one of its mufcles; the other towards the
fpine, or its nerves: it was likewife very advantageous that one of thefe conductors, which
the author diftinguifhes by the names of *conductor of the nerves*, and *conductor of the mufcles*,
and preferably the latter, fhould have a free
communication with the floor. It was in this
fituation efpecially, that the legs of the frog,
prepared as above defcribed, received violent
fhocks,

fhocks, fprang up and contracted with vivacity at each fpark drawn from the conductor of the machine, although it was at a confiderable diftance, and although the difcharge was made neither on the conductor of the nerves, nor on that of the mufcles, but on any other body, equally remote from them, and having any other communication through which the difcharge might be tranfmitted, for inftance, on a perfon placed in the oppofite corner of the room.

(2.) This phenomenon furprized Mr. Galvani, perhaps more than it ought to have done; for the power, not only of electric fparks when they immediately ftrike the mufcles or nerves of an animal, but of a current of this fluid traverfing them, in any manner whatever, with fufficient rapidity, its great power, I fay, of exciting commotions, was a thing fufficiently known; befides, it was obvious how, in this experiment, and in all thofe of the fame kind, related in the firft and fecond parts of his work, and which are reprefented in the two firft plates of figures, his frog became liable to be affected by fuch a current. We have only to confider that well-known property of electrical atmofpheres, or what is called *compreſſive electricity*, by which the fluid of conducting bodies,

placed

placed within the fphere of action of an electri-
fied body, is compreffed and difplaced, in proportion to the force and extent of this fphere, and
kept in this ftate of difplacement fo long as the
electricity fubfifts in the predominant body; and
when this is removed, returns to its place gradually, if the electricity of that body is flowly
diffipated, or in an inftant if it be deftroyed in-
ftantaneoufly, by difcharging fuddenly the body
that contained it. It is this returning current,
therefore, this reflux of electrical fluid in the
conducting bodies contiguous to the frog, or
near it, its fudden paffage from the conductor
of the mufcles to the conductor of the nerves,
or *vice verfa*, through its body, efpecially when
fuch a current is compreffed in the fingle and
narrow channel of the nerves, which excites the
fpafms and movements in the experiments in
queftion. Mr. Galvani, who feems not to have
fufficiently reflected on this property of electrical
atmofpheres, and who was not aware of the pro-
digious fenfibility of his frog, fingularly prepared in the manner above defcribed (I muft
here obferve that I have found this fenfibility
nearly equal in all the other fmall animals, fuch
as lizards, falamanders, and mice) was extremely ftruck with fuch an effect, which will
probably

probably not appear fo marvellous to other phi-
lofophers. This, however, was the firft ftep
which led him to the grand and beautiful difcovery of an animal electricity, properly fo called, and which belongs not only to frogs and
other animals of cold blood, but likewife to
every animal of warm blood, quadrupeds,
birds, &c.; a difcovery which forms the fubject
of the third part of his book, a fubject altogether new, and very interefting. It is thus
he has opened to us an immenfe field, into
which I propofe to enter, and purfue my refearches, after I fhall have dwelt a little more
on thofe preliminary experiments which relate
to the action of artificial or extraneous electricity
on the nervous and mufcular fibres.

(3.) It was chance that prefented to Mr.
Galvani the phenomenon I have been defcribing, and which aftonifhed him (I repeat it)
more than it ought to have done. Still who
would have believed that a ftream of electricity,
fo feeble as not to be rendered fenfible by the
moft delicate electrometer, fhould be capable
of affecting fo powerfully the organs of an animal, and of exciting in its limbs, cut off one
or more hours before, movements, nowife inferior in ftrength to thofe produced in the living
animal,

animal, fuch as vigoroufly darting out its legs,
fpringing up, &c. to fay nothing of the moft
violent tonic convulfions ? And yet fuch is the
ftream that affects the little animal, placed, for
inftance, on a table, near fome metal, or between two good conductors, not infulated, when
a perfon draws from the prime conductor, fufpended feveral feet above, a moderate fpark,
and conveys the difcharge through quite another
channel.

(4.) I fay *moderate*; for if it be very ftrong,
and the conductor, large and highly charged, be
not at a very confiderable diftance from the bodies on the table, little fparks will be perceptible
in the interftices of thefe bodies, efpecially the
metallic ones, and even in the place where the
frog forms a ring of communication between
them, which fparks are evidently produced by
the returning ftream of electricity, of which I
· have already fpoken, (fect. 2.) Or if matters
do not come to this point, inftead of fparks we
may perceive movements, fufficiently obvious,
of electrometers placed on the fame table and
in the fame places. In this cafe, therefore,
where the electrometer affords the fign, and
much more in the other, where the above mentioned fparks are obtained, we may obferve,
that

that even a frog, entire and untouched, or any
other small animal, as a lizard, a mouse, or a
sparrow, is feized with strong convulsions in all
its limbs, especially in its legs, which·dart for-
wards with vivacity, if the passage of the electric fluid (the returning stream) follows the direction of these same legs from one end to the
other. So far there is nothing wonderful; the
circumstance that may excite surprise is in the
case where the stream of electricity, though no
longer sensible, not even to the most delicate
electrometer, continues to excite the same convulsions, the same movements, if not in the entire frog, at least in its limbs, when dissected and
prepared in the manner practised by Mr. Galvani.

(5.) I have endeavoured, with much attention, to determine what might be the least electrical power requisite to produce these effects,
as well in the entire and living frog, as in one
dissected and prepared in the manner above described, which is what Mr. Galvani has omitted
to do. I have preferred the frog to every other
animal, because it is endowed with a very durable vitality, and it is very easy to prepare it.
I have, however, made experiments on other
small animals with the same view, and with a
success nearly similar. In order to estimate well

the

the ftrength of the ftream of electricity, I have thought it right to fubmit the animal intended for experiments of this kind, not to the returning ftreams occafioned by electrical atmofpheres (Sect. 2.), but to direct electrical difcharges, fometimes from a fimple conductor, fometimes from a Leyden phial, and in fuch a manner that the whole ftream muft have paffed through the body of the animal. For this purpofe I was careful to keep it infulated in one way or other, and moft frequently by fixing it, with pins, to two flat pieces of foft wood, fupported by glafs columns.

(6.) I have found then, that for the living and entire frog the electricity of a fimple conductor, of a middling fize, is fufficient, when it comes only to be able to give a very weak fpark, and to raife Henley's electrometer from five to fix degrees; that if I make ufe of a Leyden phial, likewife of a middling fize, a much weaker charge of this produces the effect, fuch a one, for example, as yielding not the leaft fpark, and being nowife fenfible to the quadrant-electrometer, is fcarcely fufficiently fo to Cavallo's electrometer to feparate its little pendula about 1-tenth of an inch.

(7.) This, as I have juft now fhown, for a

2   frog

frog entire and untouched; for when it is diffected and prepared in different ways, and particularly after Galvani's manner, in which the
legs are connected with the dorfal fpine merely
by the crural nerves, a much weaker degree of
electricity, whether from the conductor or from
the Leyden phial (the fluid being obliged to
pafs through the narrow paffage of the nerves),
fails not to excite convulfions, &c. Yes, an
electricity forty or fifty times weaker, as a
charge of the phial that is abfolutely impercep-
tible to the laft-mentioned electrometer (Cavallo's), and even to that extremely delicate
one of Bennet; a charge, that I was able to
render fenfible only by mears of my condenfer,
and which I think may be eftimated at five or
fix hundredths of a degree of Cavallo's electrometer.

(8.) Thus then, in the legs of a frog attached
to the fpine of the back folely by its nerves
(thefe being laid bare), we have a new fpecies
of electrometer; fince electrical charges, which
from their yielding no fign to the electrometers
already in ufe, would feem null, afford fuch obvious ones to this *animal electrometer*, if I may
be allowed the expreffion.

(9.) When

(9.) When we have feen how, in a frog
thus prepared, ftrong convulfions are excited
by an extremely weak electricity, by an imper-
ceptible ftream of fluid, we ought furely to be
no longer furprifed, that the animal fhould be
affected in the fame manner when any body
whatever difcharges fuddenly the prime conductor of an electrical machine, and occafions another ftream of electric fluid, great or
fmall, of the fluid before difplaced in the conducting bodies near the frog, and which re-
eftablifhes itfelf, in the manner already explained (Sect. 2.), to pafs rapidly through its
nerves. Let us fuppofe this returning ftream to
be fcarcely equal to that which a conductor, fufficiently bulky, throws off directly, with an
electricity that yields no fpark, and that is almoft infenfible even to Cavallo's electrometer,
or a fmall Leyden phial, charged fcarcely a tenth
of a degree of this fame electrometer; let us
fuppofe, I fay, that the ftream of electricity is
not ftronger than this, ftill it will be fufficient,
as my experiments, above related (Sect. 6.
and 7.), fhow, to excite the movements in
queftion.

(10.) But if, after the experiments juft now
referred

referred to, we ought no longer to be furprifed at thofe of Mr. Galvani, defcribed in the firft and fecond parts of his work, how can we avoid being fo at thofe entirely new and wonderful ones related in the third? Experiments in which he obtained the fame convulfions and violent movements of the limbs, without having re-courfe to any artificial electricity, or extraneous excitement, by the fimple application of a conductor, one end of which was made to touch the mufcles, and the other the nerves or fpine of the frog prepared in the manner al-ready defcribed. This conductor, he found, might be either entirely metallic, or compofed partly of metal and partly of other bodies of the clafs of conductors, as water, one or more perfons, &c. Even wood, the walls and floor of the room, might enter into the circle pro-vided they were not too dry; it was only by the interpofition of non-conducting fubftances, as glafs, rofin, and filk, that the effect was pre-vented. Bad conductors, however, did not do fo well, and only during the firft moments after the animal was prepared, and fo long as the vital powers remained in full vigour; after which good conductors only were found to fucceed,

ceed, and in a fhort time it was found impoffible to produce the effect unlefs with excellent
conductors, that is, with conductors entirely
metallic. He moreover found a great advantage
from applying a fort of metallic armour, or
coating, to that portion of the fpine which he
left attached to the crural nerves, and to the
nerves themfelves, and particularly from cover-
ing this part with a thin leaf of tin or lead.

(11.) Mr. Galvani did not confine himfelf,
in thefe truly aftonifhing experiments, to frogs ;
he extended his trials with fuccefs, not only to
feveral other animals of cold blood, but likewife to quadrupeds and birds ; in all of which
he obtained the fame refults, by means of the
fame preparations, which confifted in laying bare
fome principal nerve at the part where it paffes
into a limb fufceptible of motion, and after
arming the nerve with fome metallic fubftance,
forming a communication, by means of his conductor, between this coating and the mufcles
to which the nerve is diftributed.

(12.) It was thus he fortunately difcovered,
and demonftrated to us, in the moft evident
manner, the exiftence of a real *animal electricity*
in all, or almoft all animals. It feems in fact to
be proved by his experiments, that the electric
fluid

fluid tends inceffantly to pafs from one part to another of a living organized body, and even of detached limbs, fo long as any remains of vitality fubfift in them; that it tends to pafs from the nerves to the mufcles, or *vice verfa*, and that the mufcular movements are owing to a fimilar transfufion, more or lefs rapid. In truth, it would feem that no objections can be raifed to this, or to the manner in which Mr. Galvani explains it, by a kind of difcharge fimilar to that of the Leyden phial. But a great number of new experiments that I have made on this fubject, will ferve to fhow that many reftrictions muft be made with regard both to the thing itfelf, and to the deductions the author has drawn from it; my experiments likewife will be found confiderably to extend the phenomena attributed to this animal electricity, and will difplay it to us under a great number of new circumftances and combinations.

(13.) Mr. Galvani, purfuing the idea he has formed to himfelf from his experiments, and adhering in every refpect to the fuppofed analogy of the Leyden phial, and his conductor, imagines there is naturally an excefs of electric fluid in the nerve, or in the interior part of the mufcle, and a correfponding defect of this

fluid

fluid in the outer part, and *vice verfa*; and he
fuppofes confequently that one end of this conductor muſt communicate with the nerve, which
he confiders as the conducting wire or hook of
the phial; and the other end with the external
furface of the mufcle. All the figures of his
third and fourth plates, and all his explanations
relate to this. But if he had a little varied his
experiments, as I have done, he would have
feen that this double contact of nerve and of
mufcle, this circuit which he imagines, is not
always neceſſary. He would have found, as I
have, that the fame convulfions, the fame movements may be excited in the legs and other
limbs of frogs, and of every other animal, by
placing metallic fubftances in contact with two
parts of a nerve only, or with two mufcles, or
even with different parts of a fingle and fimple
mufcle.

(14.) It is true we are very far from fucceeding fo well in this way as in the other, and that
in this cafe it is neceſſary to have recourfe to an
artifice, of which we fhall have occafion to
fpeak more fully hereafter, and which confifts
in employing two different metals; an artifice,
which is not abfolutely neceſſary when the experiment is conducted according to Galvani's
method

method above defcribed (Sect. 10 and 11), at
leaft fo long as vitality remains in full vigour in
the animal, or in its detached limbs ; but, at
any rate, fince by arming the nerves only, or
the mufcles only, with different metals, we are
able to excite contractions in the latter, and
movements in the limbs, we muft conclude that
if there are cafes (and this may perhaps ftill be
very doubtful) where the pretended difcharge
between nerve and mufcle (Sect. 12 and 13.) is
the caufe of the mufcular movements, there are
likewife many and more frequent circumftances,
where the fame movements are obtained by
quite another play, quite another circulation of
the electric fluid.

(15.) Yes, it is quite another play of the
electric fluid, of which we may be faid rather to
difturb than to reftore the equilibrium, info-
much as it paffes from one part to another of a
nerve, a mufcle, &c. as well internally by their
conducting fibres, as externally by means of the
metallic conductors that are applied, not in
confequence of any refpective excefs or defect,
but by a peculiar action of thefe fame metals,
when they are of different kinds. It is thus I
have difcovered a new law, which is not fo
much a law of animal electricity as a law of

common electricity; to which we must attribute the most part of the phenomena, which, from the experiments of Galvani, and from several others which I made myself, seemed to belong to a true spontaneous animal electricity, but which in truth do not : they are really the effects of a very feeble artificial electricity, which is excited in a way never before suspected, by the simple application of two coatings of different metals, as I have already hinted, and which I shall explain better elsewhere.

(16.) I think it right here to say, that at the discovery of this new law, of this, till now, unknown artificial electricity, I was mistrustful of every thing that seemed to me to demonstrate a natural electricity, in the strict sense of the term, and that I was on the point of giving up this idea. But upon carefully reconsidering all the phenomena, and repeating the experiments under this new point of view, I found that some of them support such an idea, (those, for instance, in which there is no need of different coatings, or even of any coating, a simple metallic wire, or any other conducting body, performing the office of conductor between the nerve, and one of the muscles connected with it, being capable of exciting convulsions in the latter),

latter), (Sect. 10, &c.) and that thus a natural animal and properly organic electricity subsists, and cannot be entirely overturned. The phenomena which establish it, although much more limited, are however sufficient to demonstrate its existence, as I have just now mentioned, and as will more clearly be shown hereafter.

(17.) What will perhaps be found more disagreeable is, that we must likewise confine within narrower limits its influence in the animal œconomy, and give up the finest ideas we had formed of it, and which seemed to be about leading us clearly to explain muscular motion. My experiments, varied in every manner possible, show that the motion of the electric fluid excited in organs, does not act immediately on the muscles ; that it does nothing more than excite the nerves, and that the latter, put into action, excite in their turn the muscles. What this action of the nerves is ; how it propagates itself from one part of a nerve to another ; how it passes to the muscles, and how the motion of the latter results from it ; these are problems, in the explanation of which we are not farther advanced than before the discovery in question.

(18.) I come now to the experiments that prove all the assertions I have advanced in these

N 2

last

laft paragraphs. From a great number I fhall
felect only a few, which feem to me the beft
calculated to eftablifh certain principles, for the
moft part new and different from thofe adopted
by Mr. Galvani. But I muft firft fay a few words
more concerning the experiments of this writer.
I know not whether he has made others, but
thofe he has defcribed in his work are included
in too narrow a circle; in all of them the object is to lay bare and infulate the nerves, and
to eftablifh a communication, by means of conducting bodies, between thefe nerves and the
mufcles that are dependent on them (as may be
feen in all the figures of the four plates annexed
to his work), in order to excite convulfions and
movements of the mufcles, by the action of the
electric fluid. He fuppofes therefore, in every
cafe, and he explains himfelf pretty clearly on
this point, that the transfufion of the electric
fluid that is produced, whether by artificial electricity, or by natural animal electricity, muft
take place from the nerves to the mufcles, or
vice verfa; that thefe two limits at leaft muft
be included in order for the mufcular movements to take place; and in truth all the experiments he has defcribed feem to prove this.
But then they are confined, as I have juft now

I   faid,

faid, within a circle that is too limited, and beyond which he has never, or fcarcely ever, extended his inquiries. By varying the experiments of this kind in different ways, I have
fhown, that neither the one nor the other of
thofe conditions, viz. the laying bare and in-
fulating the nerves, and the touching fimulta-
neoufly thefe and the mufcles, in order to procure the fuppofed difcharge, are abfolutely neceffary (Sect. 13.). It is fufficient, when, for
inftance, we have laid bare the ifchiatic nerve of
a dog, lamb, &c. if we pafs a ftream of electricity from one part of this nerve to the other,
even though it be near, and leave all the reft
untouched and free ; it is fufficient, I fay, to do
this in order to excite in the limb very ftrong
convulfions and movements ; and this whether
we employ an extraneous artificial electricity, or
excite the electric fluid that is inherent in the
nerve itfelf. Here is the manner in which I
make thefe experiments.

(19.) EXPERIMENT A. I comprefs, with a
pair of forceps, the ifchiatic nerve a little above
its infertion into the thigh, and I apply, a few
lines higher up, a piece of money, or a plate of
metal, on this fame nerve, carefully feparated
from the parts that adhere to it, and fupported

by

by a thread, a plate of glafs, a ftick of fealing wax, a piece of dry wood, or any other fubftance that is a bad conductor. Then placing the belly of a Leyden phial, very weakly charged, on the forceps; I bring the hook into contact with the other piece of metal ; and the moment the difcharge takes place, although it be too feeble to produce the leaft fpark, convulfions take place in all the mufcles of the thigh and leg, the whole limb being agitated and fpringing up with more or lefs violence. And yet the whole of this leg, and even a part of the nerve which paffes to it, are, as we fee, out of the track which the electric fluid takes in its paffage, fo that only a fmall portion of the nerve can have been irritated ; and yet this is fufficient to occafion the convulfion of the mufcles.

(20.) EXPERIMENT B. The fame effects, that is to fay, fimilar convulfions and motions of the leg, take place, without our having recourfe to an extraneous electricity, by the difcharge which takes place, in a certain manner naturally, when we apply, as above, the fame forceps, or a plate of filver, to one part of the nerve, and a plate of fome other metal, and above all, of tin or lead, to another part, and then bring about a fimple communication between

tween them, either by an immediate contact, or by the interposition of a third piece of metal made to perform the office of a conductor.

(21.) Thus we see that the same effects, that is, convulsions and violent muscular contractions, take place without any discharge of electric fluid between the nerves and muscles, in the manner Mr. Galvani supposes; and without requiring one end of a conductor to communicate with the one, and the other end with the other. Neither is the other condition, that of laying bare the nerve, and freeing it of its adhesions, at all more necessary, as will appear from the following experiments.

EXPERIMENT C. I apply coatings, or plates, of different metals, (and it is this difference of coatings that is essential) (Sect. 14. and 15.) to an entire and living frog, that is covered with its skin, and, in short, is untouched. I apply, for example, a thin piece of tin-foil on its back, or its loins, and I place a piece of silver money under its thighs, or its belly, slightly compressing it; this done, I slide forward the piece of money till it comes into contact with the tinfoil, or I form a communication between the two metals by means of a piece of iron wire, or any other metal; and at that instant convulsive motions

take

take place in all the mufcles of the belly, thighs, and back, with violent tremors of the legs, contraction and curvature of the fpine, &c. which convulfions and fpafms, although nearly univerfal, are however moft confiderable in the limbs and mufcles contiguous to the coatings, and ftill more fo in thofe which are dependent on the principal nerves neareft to the two metals.

(22.) Thefe experiments fucceed in fome other animals; in fifhes, and particularly in eels, in none of which is it neceffary to remove the fkin, though it does not fail, in a fmall degree, to leffen the effect. This is why, by removing it, at leaft in part, particularly in the frog, we obtain the effects with more certainty, and to a greater degree. We likewife gain fomething, in this refpect, by cutting off the head of the frog, and thrufting a large pin into the fpinal marrow; we then excite, by means of different coatings in the manner above defcribed, ftronger movements, or at leaft fuch as are more obvious, becaufe they are no longer confounded with the movements the animal gives itfelf while living.

(23.) If it be advantageous, as we have feen, to take off the fkin of frogs, although very thin

and

and pretty moift, it is much more fo, and even neceffary, to remove it from almoft all the other animals, as lizards, falamanders, ferpents, tortoifes, and more efpecially from quadrupeds and birds, that are furnifhed with a drier and much thicker fkin, to fucceed in thefe experiments. The following, therefore, is the mode I adopt.

EXPERIMENT D. I faften to a table, by means of fome large pins, a lizard, a moufe, a fowl, &c. and after making an incifion through the fkin, and other integuments, to the bare flefh, on the back of the animal, I turn back the integuments on each fide; I do the fame on the thigh or the leg; after which I apply the two metallic coatings on the expofed parts, viz. on one the tin foil, and on the other a fpoon or a piece of money; I then form a communication between the two coatings, and every time I do this I excite ftrong contractions in the adjacent mufcles, and particularly in thofe of the thigh and leg, which moves and agitates itfelf with great violence. Thefe convulfions are much more confiderable when the tin foil is applied near the ifchiatic nerve, and the piece of filver on the gluteus mufcle, or on that named gaftrocnemius; and the effects are ftill greater if the nerve itfelf is laid bare, and coated with

the

the tin foil; if, leaving it attached only to the mufcles to which it is diftributed, we deprive it of every other adherent part; or if, in fhort, we feparate the entire limb from the reft of the body, with its nerve hanging out, and fubmit it in this ftate to our experiments.

I am, &c.

A. VOLTA.

September 13, 1792.

SECOND LETTER.

(24.) It will be fufficiently underftood that what I have faid with refpect to the ifchiatic nerve, and the leg, is applicable to the brachial nerve and to the arm, as well as to every other nerve relatively to the mufcles under the influence of that nerve.

(25.) Thefe laft preparations are analogous to thofe of Mr. Galvani; and they clearly prove that it is advantageous to lay bare the nerves, and ftill more fo, to detach them all round from the adherent parts; but they are far from fhow-
ing

ing that this is a neceffary condition, fince we
never fail to obtain the fame convulfions and
movements of the limbs when we fimply lay
bare the mufcles, and leave the nerves covered
and concealed under them in their natural ftate,
as all my other experiments above related (Sect.
21, 22, 23.) ferve to fhow.

(26.) After thefe trials on reptiles, birds, and
fmall quadrupeds, I proceeded to other and
larger animals, as rabbits, dogs, lambs, and
bullocks ; and I not only fucceeded in obtaining
fimilar effects in all the ways above defcribed,
but even ftronger and more durable ones, by
reafon that the vital heat maintained itfelf in
thofe large animals, and in their limbs, a longer
time. For I ought not to omit to fay, that if
in the moft part of animals of cold blood, and
particularly in frogs, the vital principle fubfifts
in detached limbs feveral hours, that principle
which renders them fo fenfible to the weakeft
electrical irritation, it hardly continues beyond
a few minutes in animals of warm blood, and
commonly difappears before the whole of this
animal heat is diffipated.

(27.) Having had fuch fuccefs with my experiments on large and fmall animals of every
kind, in fome inftances alive and entire; in

others

others deprived of their ſkin, or their head, or diſſected in different ways ; and having obtained ſimilar effects in their large detached limbs, and almoſt always without the preparation required by Mr. Galvani, that is to ſay, without laying bare the nerves, I was deſirous of going ſtill farther, and of making ſimilar trials on ſmaller limbs, on a ſingle muſcle, and even on ſmall portions of muſcles ; and the freſh ſucceſs I had in theſe trials led me to other diſcoveries, which I will ſoon mention, after having deſcribed ſome of theſe experiments.

(28.) EXPERIMENT E. I cut off, in ſome inſtances, the leg and thigh of a frog, in others, the leg only, and in ſome half or a quarter of a leg ; and on applying, as uſual, to one part of the amputated portion the tin foil, and to the other the plate of ſilver, and forming a communication between theſe two coatings, I conſtantly excited convulſions and movements; I have even ſeparated a ſingle muſcle, for inſtance the gluteus, or the gaſtrocnemius, and ſometimes only a portion of muſcle not larger than a barley-corn, and yet the ſame effects, that is to ſay, very ſtrong contractions of theſe muſcles, or parts of muſcles, have been produced by means of two different coatings, &c.

EXPE-

EXPERIMENT F. I have repeated the same
experiments on a leg, on a half or a third part
of the leg, on a single muscle, or part of a
muscle, of a fowl and other birds; on a slice of
the gluteus of a rabbit, a lamb, &c. and I have
had the same effects as long as the flesh preserved
a sensible heat. (Sect. 26.)

(29.) Thus then we are able to excite very
strong contractions in the muscles of animals of
warm as well as of cold blood, and in every
detached portion of muscular flesh; and this by
means of the simple artifice of different metallic armours or coatings, applied to the muscle
itself, without any preparation of the nerves,
and even without laying them bare. We have
besides seen that we can excite these effects quite as well, and by the same means of
metallic coatings applied to two neighbouring
parts of the same nerve, (Sect. 19, and 20. Experiments A. and B.) whence I have reason to
conclude, that there is no necessity for a discharge of electric fluid to take place between
nerve and muscle, or for any transmission of it
from the interior to the exterior part of the latter by means of the nerve and metallic conductor, as Mr. Galvani supposes, or *vice versa*:
and that there is no comparison to be made between

tween the mufcle and the Leyden phial and its
difcharge, in the experiments in queftion. In
fact, what refemblance or analogy is there to
the Leyden phial, where the two plates of metal, a communication between which is formed
by the conductor, are applied very near to each
other on the external furface of the fame nerve,
(Experiments A. and B.) or on the external furface of two mufcles, or even of the fame mufcle
(Experiments C. D. E. F.); it muft be con-
feffed it would be in vain to attempt to fupport
any analogy between any of thefe experiments
and the Leyden phial.

(30.) Experiment G. Having placed two
coatings, one of filver leaf, the other of tin
foil, on exactly correfponding parts of the two
thighs of a frog, I excited contractions of the
mufcles, and the ufual motions of the legs, at
the inftant I formed a communication between
the two coatings by means of the conductor.

(31.) Is it thus, I afk, that the difcharge of
two Leyden phials takes place, by forming a
communication between their homologous furfaces? Let us lay afide, therefore, thefe ideas
of phial and difcharge, and every forced explanation, and let us fay fimply, that in thefe and
other analogous experiments, a tranfmiffion of

the

the electric fluid takes place from one to another of two parts properly coated; a tranfmiffion determined, not by a *relative excefs* of this fluid, which cannot naturally be fuppofed between parts that are fimilar, but by the diverfity of thefe fame coatings, which muft be of different metals, as I have taken care already to point out, (Sect. 20, and 21. Experiments B. and C.) and uniformly to inculcate in the fubfequent parts of my paper.

In fact,

(32.) EXPERIMENT H. If two mufcles, or two parts of the fame mufcle, are fimilarly coated, that is, with two plates of the fame metal, both of them equal in temper and hardnefs, in foftnefs or rigidity, in the roughnefs or fmoothnefs of their furface, and both are applied in the fame manner, it will be to no purpofe to bring about a communication between them by means of a conductor, as no convulfion, no motion will take place.

(33.) I confefs it is not eafy to conceive how and why the fimple application of two diffimilar coatings, I mean of two different metals, to fimilar parts of the animal, and even to two parts very near to each other of any one mufcle, fhall difturb the equilibrium of the electric fluid,

fluid, and drawing it from its ſtate of repoſe and inactivity, ſhall induce it to paſs inceſſantly from one part to another ; which transflux takes place as ſoon as a communication, by means of the conductor, is formed between theſe two diſfimilar coatings, and continues all the time this communication ſubſiſts. But conceivable or not, and whatever may be the cauſe, it is a fact that the experiments I have already related ſufficiently prove, and which will be confirmed by many others, to the deſcription of which I ſhall endeavour to add ſome explanation. It is a fact, to be added to what we already know in electricity ; a fact which muſt ſurely appear extraordinary, and difficult to be reconciled with the laws commonly eſtabliſhed. It is truly a new and very ſingular law, which I have diſcovered ; a law that belongs not properly to animal electricity, but to common electricity, ſince this transflux of the electric fluid, a transflux, not momentary, as a diſcharge would be, but which continues as long as the communication between the two coatings ſubſiſts, and takes place whether theſe coatings are applied to living or dead animal ſubſtances, or to other conductors not metallic, but ſufficiently good, as water, or moiſt bodies. But before I proceed to the ex-
                                        periments

periments which decifively prove all that I have advanced, I think it right to offer a few more remarks on thofe I have already defcribed (Sect. 20—32.).

(34.) It would feem from thefe that by means of the fimple artifice of coatings of different metals fuitably applied, we are able to excite very ftrong convulfions in every mufcle of every animal, fo long as it continues to poffefs any degree of vitality. Such a conclufion, however, would be too general, my experiments having taught me that it is to be admitted only with certain reftrictions, as well with refpect to the claffes and genera of animals, as with refpect to the different mufcles of each animal.

(35.) And firft with refpect to the different claffes of animals ; although it has uniformly happened that all the quadrupeds, birds, fifhes, reptiles, and amphibious animals, which have been fubmitted to my experiments, exhibited the phenomena above defcribed, it is no lefs certain that worms in general, and feveral fpecies of infects, remained unaffected. I have in vain tried with worms, leeches, fnails, oyfters, and different caterpillars ; I have not even been able to excite the leaft motion in them by fmall and moderate fparks, and difcharges of artificial electricity. Here is the manner in which I proceeded.

EXPERIMENT I. I applied the tin foil, and silver leaf, to different parts, as well external as internal, of thefe fnails, leeches, earth worms, &c. and in the beft way I was able; I then formed a communication between thefe metallic coatings, fometimes by bringing them into contact with each other, and at others by means of another metal that performed the office of a conductor; but by neither of thefe means could I ever obtain the leaft motion in any part of the body.

EXPERIMENT L. I conveyed through their bodies, both when infulated and not infulated, difcharges of a Leyden phial of fufficient ftrength to excite a moderate fpark, and to give me a flight fhock, but they were not fenfibly affected by it; no motions or convulfions were produced.

(36.) Does it follow from hence that the more imperfect animals, the whole clafs of worms, and feveral fpecies of infects, are deftitute of that fenfibility and irritability, that electrical mobility, if I may be allowed the expreffion, with which other more perfect animals are endowed? I am unwilling to draw this general conclufion from my experiments, becaufe I have

as

as yet extended them only to a fmall number of worms and infects; and with regard to the latter, I think it right to obferve that I have fucceeded, without much difficulty, with craw fifh, beetles, grafshoppers, butterflies, and flies. It may not be ufelefs that I explain one of the ways in which I fucceed with thefe animals, as they are with difficulty fubmitted to experiments, on account of their minutenefs, or of the fcales with which they are covered.

EXPERIMENT M. After cutting off the head of a fly, a butterfly, beetle, &c. I flit open, with a penknife or fmall fciffars, the whole length of the corflet, and introduce deep into the flit, near the neck, a bit of tin foil, (what is improperly called filver paper is very fit for this purpofe) and a little below I introduce, and likewife deep into the flit, a bit of filver plate, or fmall filver coin; and when I bring the latter into contact with the piece of tin foil, the legs begin to bend and tremble, and the other parts, and even the trunk of the animal, are thrown into agitation. It is very amufing to excite in this manner the chirping of a grafshopper, &c.

(37.) After what I have juft now faid, I fhould be wrong to rank infects among the animals that are deftitute (like the clafs of worms

above

above mentioned) of the electrical property in
question. At the utmoft, if caterpillars appear
to be fo, it may be faid that in this ftate of lar-
va, before they have attained, by their meta-
morphofis, a perfect ftate, and acquired new
organs, &c. they may be compared in many
refpects to worms, and, like thefe, are not endowed with electric fenfibility.

(38.) In fhort, if I may be allowed to ftate
here what I think, thofe animals only that have
very diftinct limbs, with joints, and mufcles
fitted for the motion of thofe joints, or, in other
words, mufcles that are called flexors, or le-
vators, and nerves proper to regulate them, fuch
animals only, I fay, are fenfible to, and become
feized with real fpafmodic contractions in confequence of either fmall difcharges of artificial
electricity, or a weak current of fluid occafioned fimply by different metallic coatings;
which contractions and fpafms bring on the motion, and even a violent agitation of the faid
limbs. On the contrary, worms, and fuch infects as have not fufficiently diftinct limbs, or
joints properly fo called, or which are deftitute
of flexor mufcles, or enjoy only a vermicular
motion, are nowife affected by fuch an electricity. The motions of thefe animals depend on
a different

a different animal œconomy; on a different mechanifm, which in feveral fpecies has been very
well difcovered and explained. Such are my
ideas, ftill indeed fomewhat vague, and founded
only on a few experiments; it is the fequel of
thefe that muft either confirm or rectify them.

(39.) With refpect to different mufcles in the
fame animal, I am able to advance fomething
more certain. I fay then, that all mufcles are
very far from being fufceptible of contraction
from the weak electricity in queftion. There is
a great diftinction to be made with regard to
their functions in the animal œconomy; all of
them are not fubject to the empire of the will,
and fitted for fpontaneous movements: and,
ftrictly fpeaking, it is only thofe which are fo
that are capable of fpafmodic contractions by
the means above defcribed; yes, the mufcles
fubject to the will are the only ones I have found
fufceptible of irritation and motion, by the action of that weak current of electric fluid occafioned by the fimple contact of two different
metals. The other mufcles, over which the
will has no direct power, as thofe of the ftomach, inteftines, &c. are not at all fo, not even
the heart, though in other refpects fo irritable.
We muft except, however, the mufcles of the

O 3 diaphragm,

diaphragm, (and I conjectured it before I made the trial) thefe being of the number of thofe whofe motion depends on the will.

EXPERIMENT N. It is very furprifing that a flice of good mufcular flefh, cut, for inftance, from the thigh of a lamb killed half an hour or an hour before; that this piece, I fay, of mufcle, almoft quite cold, and which is no longer fenfible to the action of any mechanical or chemical ftimulus, fhould be fo powerfully affected by the electric fluid conveyed from one part of it to another, as to be feized with very ftrong fpafmodic contractions; and that, on the contrary, the heart recently taken out of the fame animal, and ftill warm and very irritable, fhould, when treated in the fame manner, with the beft adapted metallic coatings, fuffer no alteration upon our making a communication between the two metals by means of the conductor; and that its pulfations, when weakened or flackened, or altogether fufpended, fhould not be increafed, or even revived, notwithftanding all this takes place from the application of the flighteft mechanical or chemical ftimulus.

(40.) The electric fluid, therefore, which feems to be the ftimulus appropriated to the mufcles of the will, is nowife fo to the heart,

or

or to the other mufcles formed for involuntary vital and animal functions. But what will be faid if I make it appear that it is not the imme-diate or efficient caufe of motion in the volun-tary mufcles; that even in thefe it is a mediate caufe only, the nerves alone being directly af-fected by it? And yet this is what I have learned from feveral experiments; experiments that have obliged me to give up the fineft and moft extenfive ideas I had formed on the fubject. I was fond of thinking, with Mr. Galvani, that as often as a current of the electric fluid, put in motion in the organs, was impelled with a cer-tain degree of ftrength to the mufcles, this fluid did itfelf perform the office of a ftimulant, and excited the irritability which is peculiar to them; that every mufcular movement was executed in confequence of a fimilar irruption of electrical fluid into the mufcles, either by means of arti-ficial electricity, or by 'putting in motion the natural artificial electricity; that, in fhort, even the motions which are performed naturally in the living animal machine, at leaft the volun-tary motions, acknowledged the fame caufe, that is to fay, the immediate action of the elec-tric fluid on the mufcles. But I repeat it, I have found myfelf obliged, with regret, to

O 4 give

give up all thofe fine ideas by which it feemed
poffible to explain things to admiration. Yes,
we muft confiderably limit the action of electricity in animals, and confider it under another
point of view, that is to fay, as being capable
of exciting, of itfelf, the nerves, as I have already hinted, and as I fhall now proceed to
prove.

(41.) In the firft place, then, that it can act,
and that it really does act, on the nerves, and
that the latter, excited by it, excite in their turn
the mufcles connected with them, without even
the electrical ftream's arriving at thofe mufcles,
is a fact which no longer ftands in need of proofs
after thofe furnifhed by the experiments A. and
B. (Sect. 19. and 20.) and even by an experiment of Mr. Galvani, which, according to
his account, was the firft he made, and the ori-
gin of all his other experiments. It is fufficiently obvious that the electric current, in the
experiment in queftion, as well as in thofe made
by me, and which I have juft now referred to,
pervades only a part of the crural nerve, but
not one of the mufcles of the leg; and yet as
the latter depend on the nerve, they are affected with convulfions.

(42.) But I go farther, and maintain, that
even

even in the cafes where the electrical current (it will be clearly understood that I am speaking only of weak artificial discharges, or of the current which takes place by the simple application of coatings of different metals) strikes and penetrates muscles susceptible of movement, it is not by irritating the latter immediately that it occasions them to contract, but by stimulating their nerves. This is what is shown by my experiments C. and D. (Sect. 21. and 23.) where, upon the tin foil and piece of silver being applied immediately to the muscular parts of the animal, whether the animal or only a detached portion of it is the subject of the experiment, it is not so much the muscles covered by the two metallic coatings that suffer the most violent contractions, as those which depend on some principal nerve, to which one or other of the coatings is contiguous. It is in this manner that in the frog, when the tin foil is applied on the loins, where the crural nerves lay at but little depth, the muscles of the legs are seized more than any others with strong convulsions, more so even than those contiguous to the other coating, that is to say, to the piece of silver. I have already pointed out the same thing in quadrupeds, dogs, lambs, &c. with regard

gard to the ifchiatic nerve, (Experiment D.)
and I have only to add, that the leg never fails
to be convulfed when this nerve does not lay
too deep under the flefh and other integuments,
and one of the coatings is properly applied to
this part ; even although the other coating
fhould be made to correfpond neither with the
gluteus nor any mufcle of the leg, but with any
other mufcle whatever, provided it be not at too
great a diftance. Here is another proof why
this happens :

EXPERIMENT O. If we apply in a frog, or
any other fmall animal, the tin foil the whole
length of the fpine of the back, from which
proceed all the nerves of the trunk and limbs,
and the other coating to any other part whatever, all the limbs become affected ; the mufcles,
not only of the legs, but of the belly and back,
experience fpafmodic contractions, and the
trunk itfelf becomes curved ; in a word, the
convulfions are general. The experiment is
ftill more ftriking in a lizard than in a frog, and
I fhall therefore defcribe it.

EXPERIMENT P. After cutting off the head
of a lizard, and laying bare the mufcles of the
back by removing the fkin, I apply a piece of
tin foil to the mutilated end, in fuch a manner
that

that the tin foil is fpread beyond the edges of the wound, fo as to rife a little over the fhoulders, and I place a piece of money on the middle of the fpine; this done, I flide forward the piece of money till I bring it into contact with the tin foil. At that inftant the legs move, the tail twifts itfelf, and the whole body of the animal becomes agitated, and darts from right to left, and from left to right. Is not this becaufe the upper part of the fpinal marrow, the principal fource of the nerves, is irritated?

(43.) Nearly the fame effects may be obtained by a fimilar operation on a moufe, a fmall bird, &c. but in thefe it is neceffary to remove not only the fkin and other integuments, but likewife fome of the flefh, and this becaufe their back being more flefhy, the principal nerves of the fpine are more concealed by this flefh, and by the bones alfo of the vertebral tube. It is in fact eafy to comprehend that the current of electric fluid, occafioned by the two coatings, penetrating only to a certain depth the parts of the animal covered by thefe coatings, can hardly reach the fpinal marrow, or the principal branches of the nerves that enter into the interior parts of the limbs, if the bones, flefh, and other intervening integuments are of confiderable

fiderable thicknefs. The reafon alfo muft be obvious, why, in the larger animals, as dogs, lambs, &c. we fail to excite contractions in all the limbs by the application of the two coatings to the back, although ftripped of its flefh. The large trunks of the nerves remain ftill at too great a depth; and it is only the fmaller branches or ramifications that lay but a little below the coatings, and thefe branches terminate, for the moft part, only in the neighbouring external parts; confequently we fee produced only fuperficial contractions or palpitations in one or other of the mufcles : or if by chance a whole limb is put in motion, it is becaufe the nerve that goes to it, and influences this motion, is but thinly covered, fo that only a thin layer of fibres intervenes between it and one or other of the metallic coatings, as appears from Experiment D. and the following ones (Sect. 23. &c.) in which the application of one of the coatings near the ifchiatic nerve, in a dog or a lamb, was fufficient to excite confiderable movements in the leg; and the nearer the coating was to the nerve, and the thinner the layer of flefh was that furrounded it, fo much ftronger in proportion were the contractions of the limb.

2

(44.) It becomes therefore neceffary to know the fituation of the nerves, their direction, &c.; and it is requifite to remove not only the common integuments, the fat, &c. but likewife part of the flefh that covers and furrounds the nerves, in order that this furrounding mufcular fubftance may be more or lefs extenuated, previoufly to the application of the metallic coating, to enable us to obtain in the larger animals contractions in any particular limb, to fay nothing of the fuperficial contractions and palpitations of one or more mufcles. It is perhaps impoffible to excite thefe fame motions and contractions in all the limbs at once; although this is not difficult in the fmaller animals, as we have already feen, (Sect. 42. Experiments O. and P.) merely by depriving them of the fkin or a part of the other integuments; and even this is not neceffary in frogs, for in thefe animals we may leave the fkin, it being fo extremely thin and moift, as not to prevent, by its interpofition, the electrical current from reaching the principal nerves or the fpinal marrow.

(45.) But if it be neceffary to pay attention to the direction of the principal nerves, in order to bring on the contractions in the different limbs, it is not lefs fo to be careful of the pofition

fition of the coatings relatively to the mufcles;
for thofe mufcles which are neareft to one or
other of the coatings, are in general the moft
liable to contract fpafmodic convulfions, and
are oftentimes the only ones in which fuch an
effect takes place; as, for inftance, when the
coatings do not correfpond with any confiderable
nerve, or if there be a nerve, when it is furrounded with too much mufcular flefh, or is too
deeply feated.

(46.) This, and the Experiments E. F.
(Sect. 28.) where a fingle mufcle, and even a
part of a mufcle, treated in the ufual way, experienced very ftrong contractions, might lead to
a fuppofition that the electric fluid produces thefe
effects by irritating the mufcular fibres themfelves, without the intervention of nerves; the
action of which would confequently be neither
primary, nor abfolutely neceffary, as I pretend.
But an argument of this fort, founded on thefe
facts, can have no weight, unlefs it could
be proved that in thefe mufcles, or portions of
mufcles, there are no nerves; for if there are
nerves, (and certainly there muft be, and are,
nervous filaments in every fenfible portion of a
mufcle, I had almoft faid in every mufcular
fibre) I may ftill maintain that it is thefe ner-

vous filaments, ramifying through the whole
fubftance of a mufcle, that are immediately affected by the electric fluid which penetrates this
fame fubftance ; that this fluid exerting its influence on their nerves, an influence that finifhes
there, the latter exert theirs on the mufcles,
&c. I may, I fay, be able to maintain, with
fufficient probability, that the electric fluid has
no other influence, in the phenomenon of mufcular contractions, than that of exciting the
nerves ; in a word, that it is not the immediate
caufe. Such an affertion, which the things already explained render more than probable, is
proved directly, and in the moft obvious manner, by feveral experiments I have made on the
tongue ; experiments that have led me to other
difcoveries equally interefting and curious.

(47.) Having fucceeded in exciting tonic
convulfions, and the moft violent motions in the
mufcles and limbs, not only of fmall but of
large animals, without laying bare any nerves,
by the fimple application of coatings of different
metals to the mufcles when freed from their integuments, I foon thought of trying whether the
fame effects might not be obtained in the human
body. I conceived that the thing might fucceed
very well in amputated limbs ; but in the entire

tire and living fubject how was it to be effected?
It feemed likewife to be neceffary to remove the
integuments, make deep incifions, and even
diffect off portions of the flefh from the parts
on which we might think of applying the metallic coatings (as I have remarked we are often
obliged to do in the larger animals). Fortunately it came into my head, that we have, in
the tongue, a mufcle that is bare, or at leaft
deftitute of thofe thick integuments with which
the external parts of the body are covered, a
mufcle which is extremely moveable, and moveable at will. Here then, I faid to myfelf, are
all the conditions requifite to enable us to excite movements by the ufual artifice of different
metallic coatings. With this view I made, on
my own tongue, the following experiment.

(48.) EXPERIMENT Q. Having covered
the point of the tongue, and a part of its upper
furface, to the extent of fome lines, with a
piece of tin foil, (what is called filver paper is
the fitteft for the purpofe) I applied the convex
part of a filver fpoon farther on, on the flat
part of the tongue, and by inclining the fpoon
downwards brought the handle of it into contact with the tin foil. I expected to fee my
tongue affected with tremor; and on this account

count I made the experiment before a looking-
glafs. The effect, however, I had ventured to
foretel did not take place; but inftead of it I
had a fenfation I nowife expected; this was a
pretty ftrong acid tafte on the point of the
tongue.

(49.) I was at firft much furprifed at this;
but upon reflecting a little on the fact, I
eafily conceived, that the nerves which terminate on the point of the tongue, being the
nerves deftined for the fenfations of tafte, and
not for the motion of this flexible mufcle, It
was perfectly natural, that the irritation of the
electric fluid, put in motion by the ufual artifice, fhould excite a tafte, and nothing more;
and that in order to excite in the tongue the
motions of which it is fufceptible, it would be
neceffary to apply one of the metallic coatings
near its root, where the nerves enter that influence its motion; and this I foon verified by another experiment, as follows:

(50.) Experiment R. Having cut out,
from a lamb recently killed, the tongue near its
root, I applied a piece of tin foil at the end
that was cut, and the filver fpoon to one of the
furfaces of the tongue; and then forming a
communication between thefe two metallic coat-

ings, I had the pleasure to see the whole tongue affected with tremor, raising its point, and turning and bending itself in different directions, every time, and as long as such a communication took place.

(51.) I have repeated this experiment on the tongue of a calf, which I placed, coated in the same manner with a piece of tin foil near its root, on a silver plate, that the latter might serve as another coating; and the success was the same. I have likewise repeated it on the tongue of other smaller animals, as mice, chicken, rabbits, &c. and I have almost always obtained the same effect. I say *almost always*, for in the tongue of the smaller animals it sometimes failed; either because the tin foil was not applied exactly to the proper place, where the nerves that influence the motions of the tongue are inserted; or because the tongue being cold, had lost its vitality, which seldom lasts long in the muscles of animals of warm blood, as I have already had occasion to observe (Sect. 26.), and particularly in the tongue.

I am, &c.

A. Volta.

October 25, 1792.

XII. *A Return*

XIII. *An Account of a singular Case of Ischuria, in a young Woman, which continued for more than three Years; during which Time, if her Urine was not drawn off with the Catheter, she frequently voided it by vomiting; and, for the last twenty Months, passed much Gravel by the Catheter, as well as by vomiting, when the Use of that Instrument was omitted, or unsuccessfully applied. To which are added some Remarks and Physiological Observations.* By Isaac Senter, M. D. *Associate Member of the College of Physicians of Philadelphia, and senior Surgeon in the late American Army.* Vide *Transactions of the College of Physicians, of Philadelphia.* Vol. I. Part I. 8vo. Philadelphia, 1793.

THE subject of this extraordinary case was a healthy-looking servant girl, who, in June, 1785, being then in her fifteenth year, was seized with a pain in the left hypochondrium, accompanied with cough, oppression at her breast, dyspnœa, and fever.

She had menstruated pretty regularly from the age of thirteen till within five weeks of her present illness, which was ascribed to cold.

P 2   Venæ-

Venæfection and other fuitable remedies were had recourfe to by Dr. Senter, to whom fhe applied for relief, and her complaints foon fubfided; but about a month afterwards fhe vomited up a quantity of bloody pus, which induced him to think a vomica had burft in. her ftomach; for during the whole of this illnefs, her ftomach, it feems, was fo irritable, that fhe could with difficulty retain in it either food or medicine.

She had now a fuppreffion of urine, which, after continuing twenty-four hours, went off without any medical affiftance. After this fhe became regular in her menfes, and in about two months was fufficiently recovered to refume her employment as a fervant, which fhe continued to follow till the 3d of June, 1786, when all her former complaints (except the fuppreffion of the menfes) returned with greater feverity than before.

Her pulfe was now at 120; her ftomach, as during the former attack, was fo irritable, that fhe vomited up immediately almoft every thing fhe took. Of the different remedies that were had recourfe to, opium, when fhe could retain it on her ftomach, and repeated blood-letting in fmall quantities, gave her the moft relief.

On

On the 2d of July, when the feverity of the fymptoms had fubfided, fhe was feized with a total fuppreffion of urine, which continued till the beginning of the fixth day, when a vomiting came on, which lafted till fhe brought up nothing but water; and this water, fhe faid, tafted like urine.

As the vomiting continued fhe found relief from the forenefs nnd fwelling fhe had felt for feveral days in the lower part of the abdomen.

She now thought herfelf much better, but the vomiting continued to return, more or lefs, every day, till the 14th of July, when Dr. Senter again faw her, and prevailed on her to fubmit to the introduction of a catheter, by means of which he drew off about three pints of clear, but high-coloured, urine.

From this time, till December, fhe continued with very little abatement of her complaints ; and as fhe could lie in no other pofition, was conftantly fupported in an arm chair, in a reclined pofture, with pillows under her hips.

During the whole of this period, whenever her water was omitted to be drawn off once in thirty or thirty-fix hours at fartheft, fhe never failed, we are affured, to vomit it up. To af-

certain

certain fo extraordinary a fact, our author tells
us he often vifited her about the time he knew
fhe muft vomit if the catheter was not introduced; and after examining her bladder, and
finding it full, hard, and tender, fat by her till
the vomiting returned, faved the water that fhe
brought up in this way, and on comparing it
with what he drew off by means of the catheter, found it the fame in every refpect.

During the time her urine came off by vomiting, fhe fuffered, it feems, great anxiety
and thirft, and complained of a fenfation of
inverfion or turning up of fomething (running,
as fhe expreffed it) that appeared to tear her
bowels.

In January, 1787, from fome caufe unknown,
fhe could not be relieved with the inftrument,
nor could fhe vomit up her urine for feveral
days; but at length it paffed by the navel for
three days fucceffively; after which the catheter
was ufed with the fame effect as before.

About the beginning of Auguft a brick-co-
loured gravel began to pafs off through the catheter, and continued to be difcharged in confiderable quantity, whenever her urine was
drawn off, till the beginning of November; at
which time fhe felt more diftrefs than ufual,

when-

whenever her urine came off by vomiting, and
she soon obferved a gritty fubftance in her
mouth. When our author was informed of this
new phenomenon, he requefted her to fave the
urine for his infpection the next time she vomited; and on comparing it with what he drew
off, found it contained the fame kind of gravel
as that which paffed the catheter.

From this period, to the fummer of 1788,
her complaints, he obferves, continued much
the fame; but during that fummer she twice
paffed a fmall quantity of urine through the
urethra, each time in confequence of being
frightened. The hypogaftrium became more
tumid, and she complained of great forenefs
about the bladder, even after it was evacuated;
the bladder itfelf feemed to be much thickened, and the apparent inequality of its furface
was fo great, and the tumour fometimes shifted
fo towards the right or left inguen, according
as her body was moved, that our author fufpected the exiftence of a ftone.

Through the month of September her urine,
we are told, could very rarely be drawn off;
for upon the introduction of the catheter, a
fpafm feized the urethra and neck of the bladder, fo that although the inftrument feemed to

pafs

pafs high up into the bladder, not more than a gill of urine could be drawn off, before it ftopped entirely, with a fenfation of fomething falling down againft the cervix, which fhe was confident was a ftone; and early in the following month, Dr. Senter being able to introduce a found, readily met with a ftone, which feemed to be of a fmall fize, and fofter than urinary calculi commonly are.

She had at different feafons of the year feveral fmall abfceffes on different parts of her body, but they did not appear to relieve her general complaints. She alfo voided at times (after fhe had thrown up her urine) a bloody pus, of a coppery tafte. This purulent difcharge, it is obferved, was never expectorated by coughing, though fhe had at times a dry cough, but was conftantly brought up by vomiting.

In the fpring of 1789 her urine began to pafs *per anum*, loaded with the fame kind of gravel that had come away by the catheter. This diminifhed but did not put a ftop to her vomiting; for fhe continued to throw up more or lefs gravel that way every week. This new courfe of her urine occafioned a troublefome diarrhœa and tenefmus, but fhe felt lefs inconvenience from the ftone in the bladder.

After

After the 13th of May her bladder never became fo much diftended with urine as it had been before; and the fecretion of urine, as well as the formation of gravel, we are told, evidently diminifhed in proportion to her lofs of ftrength, and the increafe of the diarrhœa. The menfes, which, during the whole of her illnefs, had returned at irregular periods, now entirely ceafed. During the fummer, the frequency of vomiting increafed; fhe had feveral convulfive fits after vomiting; became more and more emaciated, and hectical; and, at laft, lethargic; and on the 11th of Auguft, 1789, died.

The body was examined the day after her death, by Dr. Senter, in the prefence of Dr. Waterhoufe, of Cambridge, and Dr. Mafon, of Philadelphia, who, as well as feveral other refpectable medical practitioners, had occafionally vifited her in her life-time, and feen her vomit up both urine and gravel.

On diffection, nothing was difcovered that could throw any light on the nature of the difeafe.

In the thorax, the only morbid appearance was an adhefion of part of the right lobe of the lungs to the pleura.

In

In the abdomen, the omentum was found much wafted, and of a dark gangrenous colour; the ftomach alfo is defcribed as being in a gangrenous ftate, and containing ' a femi-pu-' rulent matter, of a foetid fcent;' but the weather, we find, was very warm, and the body in an offenfive ftate, at the time the diffection was made. Nothing particularly worthy of notice was obferved in the ftate of the liver, gallbladder, inteftines, kidneys, or ureters. The urinary bladder was alfo in its natural ftate, not in the leaft thickened, and contained no fand or gravel. The uterus contained about a drachm of thick, foetid pus, but had no other appearance of difeafe; the Fallopian tubes were larger than ufual, and ftrung with feveral hydatids of the fize of a walnut; the corpora fimbriata had a gangrenous appearance; the ovaria were enlarged to the fize of a fmall hen's egg, and diftended with a clear limpid fluid.

To the preceding hiftory Dr. Senter has added many judicious remarks; and in his attempt to account for the phenomena of fo very uncommon a cafe, has not omitted to avail himfelf of the modern doctrine of the retrograde motion of the lymphatics, and of the opinions of thofe writers who have maintained
the

the exiftence of a direct communication between
the alimentary canal and the urinary bladder.

There are many inftances, he obferves, in
medical books, of fudden and partially-in-
creafed actions of the veffels of the human
body; but he candidly acknowledges that his
reading has furnifhed him with no fact fimilar
to the extraordinary one which is the fubject of
the paper before us * : that which he confiders

as

---

* There are, however, upon record, two cafes which exhibit a ftriking analogy to that of Dr. Senter's patient ; and
although they may have been overlooked, or perhaps difre-
garded on a fuppofition of their improbability, they muft
now become extremely interefting by the tendency they have
to corroborate the curious and extraordinary facts he has related. Both the cafes we allude to occur in the Hiftory of
the Academy of Sciences at Paris, and are as follows :

Cafe I. " M. Maraldi has communicated to the Aca-
" demy the following cafe, from a letter addreffed to him
" by M. Marangoni, phyfician at Mantua :

" A Nun, of the Order of St. Francis, in the convent of
" St. Jofeph, at Mantua, aged thirty-five years, of a thin
" and delicate habit of body, and who had long been fub-
" ject to hyfterical complaints, was attacked with pains,
" fpafms, and fwelling of the abdomen, to which fuccceded
" a violent and alarming fuppreffion of urine. Soon after
" this fhe felt a pain, which fhe defcribed as afcending from
" the lower part of the abdomen to her ftomach; and fhe
" vomited

as coming the neareſt to it, is a caſe deſcribed
by Dr. Percival, in the ſecond volume of his
Eſſays, Medical and Experimental, (8vo,
London,

" vomited a fluid which, without any difficulty, was
" known to be urine. This vomiting continued forty days,
" during which time the patient voided no urine by the
" uſual channel, unleſs the ſurgeon drew it off with a ca-
" theter, and even then the quantity ſcarcely amounted to
" an ounce a day. At the end of the forty days, the urine
" ſpontaneouſly reſumed its natural courſe, and in a day or
" two the patient found herſelf perfectly recovered. But
" the vomiting of urine returned, and at the end of twenty-
" ſeven days, the patient complained of very acute pain
" about the region of the pubis. Her ſurgeon was deſirous
" of relieving her by means of the catheter, but there was
" ſuch a contraction of the urethra, that he found it impoſ-
" ſible to introduce even a probe into the bladder. The
" vomiting of urine has continued, and what is remarka-
" ble, there is no appearance of food mixed with it, even
" when the vomiting takes place ſoon after her meals.
" When M. Marangoni wrote this account, the patient had
" been in this ſtate thirty-two days.
   " This ſingular complaint would lead one to think there
" is an immediate though hitherto undiſcovered commu-
" nication between the ſtomach and the urinary bladder;
" but M. Marangoni and the celebrated Lanciſi are of a
" different opinion; they both of them think, that in caſes
" of this kind a ſuppreſſion of urine takes place in the kid-
" neys; that is to ſay, that the kidneys ceaſe to extract this
                                                      " fluid

London, 1773) of a woman who, after a spontaneous vomiting of several days, during which she brought up three gallons of water, was entirely cured of a dropfy of the ovarium.

" fluid from the blood, and that in their ftead the glands of
" the ftomach perform this function."

Cafe II. " M. Lemery is acquainted with a Monk, who,
" for about eight years, has been fubject to a periodical vo-
" miting, the fits of which are as regular as thofe of a quar-
" tan ague. Five hours, or thereabouts, before the vomit-
" ing begins he complains of violent pains in his kidneys.
" The vomiting continues, with intervals, four or five
" hours. What he vomits is of a dirty red colour. It is
" almoft entirely water, but has a ftrong urinous fmell, and
" the patient has no doubt of its being really urine, as he
" eats but very little, and drinks more than the ufual portion
" of a Monk. He drinks only wine, the colour of which
" agrees with that of the fluid he vomits. A few hours
" after the vomiting he finds himfelf well, and remains fo
" till the next fit. He ufes a great deal of exercife, with-
" out which he thinks he fhould fuffer more. It is a known
" fact, that in nephritic pains, which are always occafioned
" by obftructions of the kidneys, the patients are fubject to
" frequent vomiting, and that what they bring up fmells
" much of urine."——*See* Hiftoire de l'Academie Royale
des Sciences, Années 1715 & 1722. EDITOR.

1. THOUGHTS on the Effects of the Application and Abstraction of Stimuli on the Human Body; with a particular View to explain the Nature and Cure of Typhus. By *J. Wood*, M. D. 8vo. *Murray*, London, 1793.

2. An Account of the Bilious, Remitting, Yellow Fever, as it appeared in the City of Philadelphia in the year 1793. By *Benjamin Rush*, M. D. 8vo. Philadelphia, 1794.

3. Observations on the Cause, Nature, and Treatment of the Epidemic Disorder prevalent in Philadelphia. By *D. Nassy*, M. D. Member of the American Philosophical Society. 8vo. Philadelphia, 1793.

4. A Short Account of the Malignant Fever, lately prevalent in Philadelphia; with a Statement of the Proceedings that took place on the Subject in different Parts of the United States. By *Matthew Carey*. 8vo. Philadelphia, 1793.

5. A Treatise on the Extraction of the Cataract. By *Frederick Bischoff*, F. M. S. Oculist to his Majesty in the Electorate of Hanover,

I   and

and to her Majefty in England. 8vo. *Nicol*, London, 1793.

6. An Account of a Fever which appeared in feveral Parts of Somerfetfhire in the year 1792. By *Richard Poole*, Surgeon, Sherborne. 8vo. *Johnfon*, London, 1793.

7. A Guide for Self-Prefervation and Parental Affection; or plain Directions for enabling People to keep themfelves and their Children free from feveral common Diforders. By *Thomas Beddoes*, M. D. 12mo. *Murray*, London, 1793.

8. A Chemical Differtation on the Thermal Waters of Pifa, and on the neighbouring acidulous Spring of Afciano; with an Hiftorical Sketch of Pifa, and a Meteorological Account of its Weather. To which are added, Analytical Papers refpecting the Sulphureous Water of Yverdun. By *John Nott*, M. D. of Briftol Hot-wells. 8vo. *Walter*, London, 1793.

9. Horti Botanici Cantabrigienfis Catalogus. 8vo. Cantabrigiæ, 1794.

10. Flora Oxonienfis, exhibens Plantas in agro Oxonienfi fponte crefcentes, fecundum Syftema fexuale diftributas. Auctore *Joanne Sibthorp*, M. D. Profeffore Regio Botanico, Regiæ Societatis Londinenfis aliarumque Societatum Socio. 8vo. Oxonii, 1794.

11. Differ-

11. Differtàtio Inauguralis de Angina maligna. Auctore *Arthuro Bedford*, Anglo. 8vo. Edinburgi, 1792.

12. Differtatio Inauguralis de Refpiratione. Auctore *Thoma Blair*, Scoto-Britanno. 8vo. Edin. 1792.

13. Differtatio Inauguralis de Variolis. Auctore *Joanne Bower*, Scoto. 8vo. Edin. 1792.

14. Differtatio Inauguralis de Vifu. Auctore *Wheaton Bradiſh*, Hiberno. 8vo. Edin. 1792.

15. Differtatio Inauguralis de Rheumatifmo acuto. Auctore *Joanne Bradley*, Anglo. 8vo. Edin. 1792.

16. Differtatio Inauguralis de Cœli Effectibus. Auctore *Jacobo Buchan*, Scoto. 8vo. Edin. 1792.

17 Differtatio Inauguralis de Rheumatifmo acuto. Auctore *Andrea Grieve*, Scoto. 8vo. Edin. 1792.

18. Differtatio Inauguralis de Hypochondriafi. Auctore *David Corbin Kerr*, Virginienfe. 8vo. Edin. 1792.

19. Differtatio Inauguralis de Variolis. Auctore *Gulielmo Marſden*, Anglo Britanno. 8vo. Edin. 1792.

20. Differtatio Inauguralis de Pneumonia.

2 Auctore

Auctore *Carolo Merivether*, Virginienfe. 8vo. Edin. 1792.

21. Differtatio Inauguralis de Variolis. Auctore *Roberto Montgomery*, Hiberno. 8vo. Edin. 1792.

22. Differtatio Inauguralis de Hydrope Anafarca. Auctore *Thoma Pollard Pierce*, Barbadenfe. 8vo. Edin. 1792.

23. Differtatio Inauguralis de Angina maligna. Auctore *Georgio Wier*, Scoto. 8vo Edin. 1792.

24. Differtatio Inauguralis de Alimento. Auctore *Gulielmo Yates*, Anglo. 8vo. Edin. 1792.

25. Differtatio Inauguralis de Coitu ejufque variis Formis quatenus Medicorum funt. Auctore *Johanne Paul Gottleib Kircheifen.* 4to. Jena, 1792.

26. Analyfe du Syfteme abforbant ou lymphatique. Par M. *des Genettes*, D. M. 8vo. Montpellier, 1791.

27. Memoire fur une Maladie de l'Ovaire. Par *Jean Baptifte Ph. R. N. Laumonier*, Chirurgien en chef de l'Hotel Dieu de Rouen. 4to. Rouen, 1790.

28. Avis aux Sages Femmes; par M. *Sacombe*,

Medecin-Accoucheur, Membre de plusieurs
Academies. 8vo. Paris, 1792.

29. Recherches Physico-chymiques. Cahiers
I. II. III. 4to. Amsterdam, 1793-4.

30. Memoires de l'Academie Royale des
Sciences et Belles Lettres depuis l'Avenement de
Frederic Guillaume II. au Trone. 1788 et
1789. Avec l'Histoire pour le même Tems.
4to. Berlin, 1793.

31. Sammlung der Deutschen Abhandlungen
welche in der Königlichen Akademie der Wif-
fenfchaften zu Berlin vorgelefen worden in den
Jahren 1788 und 1789. *i. e.* A Collection of
German Effays, read before the Royal Academy
of Sciences at Berlin, in the Years 1788 and
1789. 4to. Berlin, 1793.

32. Memoria Chirurgica ful Labbro leporino
complicato; di *Giufeppe Sonfis*, R. Affeff. Medico, &c. 4to. Cremona, 1793.

33. Pifaura Automorpha e Coreopfis formofa;
Piante nuove pubblicate da *Giufeppe Antonio Bo-
nato*, Dott. di Medicina, pubblico Bibliotera-
rio, Ifpettore e Soprantendente all' Orto medico dell' Univerfita di Padova. 4to. Padova,
1793.

INDEX.

INDEX.

I

Swietenia,

END OF THE SIXTH VOLUME.

www.ingramcontent.com/pod-product-compliance
Lightning Source LLC
Chambersburg PA
CBHW021527210326
41599CB00012B/1406